Texts in Applied Mathematics

Volume 47

More information about this series at http://www.springer.com/series/1214

Texts in Applied Mathematics

Volume 17

Editors-in-chief:

Stuart Antman, *University of Maryland, College Park, USA*
Leslie Greengard, *New York University, New York City, USA*
Philip Holmes, *Princeton University, Princeton, USA*

Series Editors:

John B. Bell, *Lawrence Berkeley National Laboratory, USA*
Joseph B. Keller, *Stanford University, Stanford, USA*
Robert Kohn, *New York University, New York City, USA*
Paul Newton, *University of Southern California, Los Angeles, USA*
Charles Peskin, *New York University, New York City, USA*
Robert Pego, *Carnegie Mellon University, Pittsburgh, USA*
Lenya Ryzhik, *Stanford University, Stanford, USA*
Amit Singer, *Princeton University, Princeton, USA*
Angela Stevens, *Universität Münster, Münster, Germany*
Andrew Stuart, *California Institute of Technology, USA*
Thomas Witelski, *Duke University, Durham, USA*
Stephen Wright, *University of Wisconsin-Madison, Madison, USA*

More information about this series at http://www.springer.com/series/1214

Hilary Ockendon • John R. Ockendon

Waves and Compressible Flow

Second Edition

 Springer

Hilary Ockendon
Mathematical Institute
University of Oxford
Oxford, UK

John R. Ockendon
Mathematical Institute
University of Oxford
Oxford, UK

ISSN 0939-2475   ISSN 2196-9949   (electronic)
Texts in Applied Mathematics
ISBN 978-1-4939-8036-9   ISBN 978-1-4939-3381-5   (eBook)
DOI 10.1007/978-1-4939-3381-5

Mathematics Subject Classification (2010): 76-02, 76Nxx, 76Bxx

Springer New York Heidelberg Dordrecht London

Printed on acid-free paper

Springer Science+Business Media LLC New York is part of Springer Science+Business Media (www.
springer.com)

Contents

Chapter 1
Introduction

This book owes its origins to a lecture course that was first conceived five decades ago and then modified in the 1970s to form the basis for "Inviscid Fluid Flows" [42]. That monograph was retitled and rewritten to reflect developments in the 1990s as the first edition of "Waves and Compressible Flow" which was published in 2002. This edition has been further expanded to show how the methodologies expounded in the earlier books can shed light on even more phenomena outside the area of fluid mechanics.

The cold war was at its height when Alan Tayler gave his first course on compressible flow in the early 1960s. Naturally, his material emphasised aeronautics, which was soon to be encompassed by aerospace engineering, and it concerned flows ranging from small-amplitude acoustics to large-amplitude nuclear explosions. The area was technologically glamorous because it described how only mathematics could give a proper understanding of the design of supersonic aircraft and missiles. It was also mathematically glamorous because the prevalence of "shock waves" in the physically relevant solutions of the equation of compressible flow led many students into a completely new appreciation of the theory of partial differential equations. Suddenly there was the challenge to find not only non-differentiable but also genuinely discontinuous solutions of the equations and the simultaneous problem of locating the discontinuity. This led to enormous theoretical developments in the theory of *weak solutions* of differential equations and more generally to the whole theory of *moving boundary problems*.

The many dramatic theoretical developments that have occurred since then in all branches of applied science are the driving force for the evolution of this book. For applied mathematicians, the computer revolution has been the most striking and enduring of these developments, and it has resulted in an ever-increasing demand for quantitative understanding from the whole scientific community. Even when attention is restricted to wave phenomena, books such as Billingham and King [5] reveal that it would take many volumes to cover the range of pertinent applications in the physical, biological and social sciences. Nonetheless, it has emerged that the methodologies that have been so successful in treating models for waves in

© Springer Science+Business Media New York 2015
H. Ockendon, J.R. Ockendon, *Waves and Compressible Flow*,
Texts in Applied Mathematics 47, DOI 10.1007/978-1-4939-3381-5_1

gases and liquids are also suitable for applications in other areas of physical science. This was reflected in the first edition of this book which contained discussions of the application of these methodologies to electromagnetism and elasticity. This edition goes further by introducing some simple models from metal plasticity and plasma physics. These show how the insights that mathematics can give into easily observable phenomena in fluid mechanics apply equally in areas in which experimental observations are much more problematic. Equally exciting are the subtle differences encapsulated in these new models resulting from mechanisms such as rate dependence and stochasticity; both of these ideas lead to fascinating new twists to the classical theory of waves in fluids. At the same time, this mathematical knowledge transfer provides invaluable quality control for the computer codes that pervade all the areas we address. Our abiding belief is that fluid mechanics provides the best possible vehicle for someone wishing to learn applied mathematical methodologies because the phenomena are at once so familiar and so fascinatingly complex.

The layout of this book is as follows. We begin in Chapter 2 with a derivation of the equations of compressible fluid flow that is as simple as possible while still being self-contained. The only prerequisite is a belief in the ideas of conservation of mass, momentum and energy together with some elementary thermodynamics. In Chapter 3 we distil the model for acoustics, which is the simplest linear wave motion model to emerge from the general equations of gas dynamics. This equation provides the basis for several other models for wave propagation in a variety of application areas, and we will see that in many cases we are led to models that take the form of linear hyperbolic partial differential equations. We recall some of the more important exact solutions of these linear models in Chapter 4 and discuss the phenomena that arise, particularly the idea of dispersion. This discussion includes the theory of waves that have a purely harmonic time-dependence, sometimes called monochromatic waves or waves in the frequency domain. In the frequency domain, the models often reduce to elliptic partial differential equations, but the questions that need to be answered are often very different from those traditionally associated with elliptic equations. Following on from this we look at high-frequency approximations in frequency domain models. This leads us to the increasingly important "ray theory" approach to wave propagation which opens up fascinating new mathematical challenges and analogies in subjects ranging from quantum mechanics to celestial mechanics.

Nonlinearity presents the most formidable theoretical challenge in the book, and we split the discussion into two parts. In Chapter 5, we consider situations where the solutions of the nonlinear problems remain smooth despite their large amplitudes as may happen, for example, near resonance for surface gravity waves or in the far field generated by linear acoustic wave patterns. Then in Chapter 6, we describe situations in which shock waves occur, and the structure of these solutions varies dramatically as we move from gas dynamics and hydraulics to plasticity and plasma physics.

The book is written, as were its predecessors, at a level that assumes that the reader already has some familiarity with basic fluid dynamics modelling. Some knowledge of partial differential equations and asymptotic analysis including

Laplace's method and the method of stationary phase is also helpful; we do not have space to give full accounts of these methods but we do give a brief overview and references to texts where the reader can find all the details. No prior knowledge of electromagnetism, plasma physics, elasticity or plasticity is necessary.

The starred sections describe more advanced topics and can be omitted at a first reading. The exercises are an integral part of the book; those marked R are recommended as they contain basic material whereas the starred exercises are either harder or refer to work in the starred sections of the text.

Both authors reiterate their great debt to their mentor Alan Tayler. They are also very grateful for the helpful comments received from John Allen, David Allwright, Paul Dellar, Max Gunzburger, Katerina Kaouri, Laura Kimpton, Robert McKay, Mauro Perego, Maxim Zyskin and, especially, Peter Howell. Last but not least we thank Margaret Sloper for her patient retyping.

Chapter 2
The Equations of Inviscid Compressible Flow

In this chapter, we will derive the equations of inviscid compressible flow of a perfect gas. We will do this mostly by making the traditional assumption that we are working on length scales for which it is reasonable to model the gas as a *continuum*, that is to say it can be described by variables that are "smoothly" defined[1] almost everywhere. This means that the gas is infinitely divisible into smaller and smaller *fluid elements* or *fluid particles*, and we will see that it will help our understanding to relate these particles to the "particles" of classical mechanics.

This approach will of course become physically inaccurate at small enough scales because all matter is composed of molecules, atoms and subatomic particles. This is particularly evident for gases especially when they are in a rarified state as, for example, is the case in the upper atmosphere. In order to treat such gases when the mean free path of the molecules is large enough to be comparable with the other length scales of interest (such as the size of a space vehicle), it is necessary to resort to the ideas of statistical mechanics. As described in Chapman and Cowling [8], this leads to the well-developed but much more difficult *kinetic theory of gases*. Fortunately, when the limit of this theory is taken, on a scale which is much greater than a mean free path, the equations which we derive below can be retrieved. Our brief discussion of models for ionised plasmas will reveal some of the rudiments of this theory.

2.1 The Field Equations

With the continuum approach, the state of a gas may be described in terms of its velocity $\mathbf{u}$, pressure p, density ρ and absolute temperature T. If the independent variables are $\mathbf{x}$ and t, where $\mathbf{x}$ is a three-dimensional spatial vector with components

---

[1] We hope the reader will not be deterred by such imprecision, which is necessary to keep applied mathematics texts reasonably concise.

© Springer Science+Business Media New York 2015
H. Ockendon, J.R. Ockendon, *Waves and Compressible Flow*,
Texts in Applied Mathematics 47, DOI 10.1007/978-1-4939-3381-5_2

either (x, y, z) or (x_1, x_2, x_3) referred to inertial Cartesian axes and t is time, then we have an *Eulerian* description of the flow. An alternative description, in which attention is focussed on a fluid particle, is obtained by using $\mathbf{a}, t$ as independent variables, where $\mathbf{a}$ is the initial position of the particle. This is a *Lagrangian* description. A particle path $\mathbf{x} = \mathbf{x}(\mathbf{a}, t)$ is obtained by integrating $\dot{\mathbf{x}} = \mathbf{u}$ with $\mathbf{x} = \mathbf{a}$ at $t = 0$, where the dot denotes differentiation with respect to t keeping $\mathbf{a}$ fixed, and this relation may be used to change from Eulerian to Lagrangian variables. The two descriptions are equivalent but for most problems the Eulerian variables are found to be more useful.[2]

It is important to distinguish between differentiation "following a fluid particle", which is denoted by d/dt, and differentiation at a fixed point, denoted by $\partial/\partial t$. If $f(\mathbf{x}, t)$ is any differentiable function of the Eulerian variables $\mathbf{x}$ and t, then

$$\frac{df}{dt} = \frac{\partial f}{\partial t} + (\mathbf{u}.\nabla)f, \qquad (2.1)$$

where ∇ is the gradient operator with respect to the $\mathbf{x}$-components. The derivative df/dt is called the *convective derivative*, and the term $(\mathbf{u}.\nabla)f$ is the convective term which takes account of the motion of the fluid.

We have already assumed that the fluid is a continuum and this implies that the transformation from $\mathbf{a}$ to $\mathbf{x}$ is, in general, a continuous mapping which is one-to-one and has an inverse. We will also restrict attention to flows for which this mapping is continuously differentiable almost everywhere. The Jacobian of the transformation, $J(\mathbf{x}, t) = \partial(x_1, x_2, x_3)/\partial(a_1, a_2, a_3)$, represents the physical dilatation of a small element following the fluid. In order to understand the evolution of a fluid flow, it will be helpful to work out how J changes following the fluid. Since the transformation from $\mathbf{a}$ to $\mathbf{x}$ is invertible and continuous, J will be bounded and non-zero and its convective derivative will be

$$\frac{dJ}{dt} = \frac{\partial(\dot{x}_1, x_2, x_3)}{\partial(a_1, a_2, a_3)} + \frac{\partial(x_1, \dot{x}_2, x_3)}{\partial(a_1, a_2, a_3)} + \frac{\partial(x_1, x_2, \dot{x}_3)}{\partial(a_1, a_2, a_3)}$$

$$= \frac{\partial(u_1, x_2, x_3)}{\partial(a_1, a_2, a_3)} + \frac{\partial(x_1, u_2, x_3)}{\partial(a_1, a_2, a_3)} + \frac{\partial(x_1, x_2, u_3)}{\partial(a_1, a_2, a_3)}.$$

Writing the first term out, we see that

$$\frac{\partial(u_1, x_2, x_3)}{\partial(a_1, a_2, a_3)} = \begin{vmatrix} \dfrac{\partial u_1}{\partial a_1} & \dfrac{\partial u_1}{\partial a_2} & \dfrac{\partial u_1}{\partial a_3} \\ \dfrac{\partial x_2}{\partial a_1} & \dfrac{\partial x_2}{\partial a_2} & \dfrac{\partial x_2}{\partial a_3} \\ \dfrac{\partial x_3}{\partial a_1} & \dfrac{\partial x_3}{\partial a_2} & \dfrac{\partial x_3}{\partial a_3} \end{vmatrix}.$$

---

[2] We make this remark in the context of understanding the *mathematical* basis of models for compressible flow. For computational fluid dynamics, particle-tracking methods are often more appropriate than discretisations based on Eulerian variables.

However,

$$\frac{\partial u_1}{\partial a_i} = \frac{\partial u_1}{\partial x_1}\frac{\partial x_1}{\partial a_i} + \frac{\partial u_1}{\partial x_2}\frac{\partial x_2}{\partial a_i} + \frac{\partial u_1}{\partial x_3}\frac{\partial x_3}{\partial a_i},$$

and so, using the properties of determinants, we obtain

$$\frac{\partial(u_1, x_2, x_3)}{\partial(a_1, a_2, a_3)} = J\frac{\partial u_1}{\partial x_1}.$$

The other two terms can be treated similarly and so

$$\frac{dJ}{dt} = J\nabla.\mathbf{u}. \tag{2.2}$$

We can now consider the rate of change of any property, such as the total mass or momentum, in a *material volume* $V(t)$, which is defined as a volume which contains the same fluid particles at all times. For any differentiable function $F(\mathbf{x})$,

$$\frac{d}{dt}\left[\int_{V(t)} F(\mathbf{x}, t)\, dV(\mathbf{x})\right] = \frac{d}{dt}\left[\int_{V(0)} F(\mathbf{x}(\mathbf{a}, t), t)J\, dV(\mathbf{a})\right]$$

$$= \int_{V(0)} \frac{d}{dt}[F(\mathbf{x}(\mathbf{a}, t), t)J]\, dV(\mathbf{a})$$

$$= \int_{V(0)} \left(\frac{dF}{dt}J + FJ\nabla.\mathbf{u}\right) dV(\mathbf{a}) \quad \text{(on using (2.2))}$$

$$= \int_{V(t)} \left(\frac{dF}{dt} + F\nabla.\mathbf{u}\right) dV(\mathbf{x}). \tag{2.3}$$

This formula for differentiating over a volume which is "moving with the fluid" is called the *Transport Theorem*. Using (2.1) and denoting the outward normal to $\delta V(t)$ by $\mathbf{n}$, we can rewrite (2.3) as

$$\frac{d}{dt}\left[\int_{V(t)} F\, dV\right] = \int_{V(t)} \left(\frac{\partial F}{\partial t} + \nabla.(F\mathbf{u})\right) dV \tag{2.4}$$

$$= \int_{V(t)} \frac{\partial F}{\partial t}\, dV + \int_{\partial V} F\mathbf{u}.\mathbf{n}\, dS, \tag{2.5}$$

on using the divergence theorem. Thus, from (2.5), the derivative can be interpreted as the sum of the term $\int_V(\partial F/\partial t)\, dV$, which would be the answer if V were fixed in space, and $\int_{\partial V} F\mathbf{u}.\mathbf{n}\, dS$ which is an extra term resulting from the movement of V. Note that (2.5) is a generalisation of the well-known formula for differentiating a one-dimensional integral

$$\frac{d}{dt}\left(\int_{a(t)}^{b(t)} f(x, t)\, dx\right) = \int_{a(t)}^{b(t)} \frac{\partial f}{\partial t}\, dx + f(b, t)\frac{db}{dt} - f(a, t)\frac{da}{dt}.$$

We also remark that the function $\mathbf{u}$ in (2.5) does not have to be the velocity of the fluid everywhere inside V because we only require that $\mathbf{u}.\mathbf{n}$ should be the velocity of the boundary of V normal to itself.

We now apply the Transport Theorem to derive the equations which govern the motion of an inviscid fluid. Conservation of the mass of any material volume $V(t)$ can be written as

$$\frac{d}{dt}\left(\int_{V(t)} \rho \, dV\right) = 0,$$

where ρ is the fluid density or, using (2.4), as

$$\int_{V(t)} \left(\frac{\partial \rho}{\partial t} + \nabla.(\rho \mathbf{u})\right) dV = 0.$$

If we now shrink V to a small neighbourhood of any point, we derive the differential equation

$$\frac{\partial \rho}{\partial t} + \nabla.(\rho \mathbf{u}) = 0. \tag{2.6}$$

This equation is known as the *continuity equation*. We must emphasise that the above argument relies crucially on the differentiability of ρ and $\mathbf{u}$. If, as will be seen to be the case in Chapter 6, the variables are integrable but not differentiable, conservation of mass will just lead to the statement that $\int_{V(t)} \rho dV$ is independent of time.

We next consider the linear momentum of the fluid contained in $V(t)$. The forces created by the surrounding fluid on this volume are the "internal" surface forces acting on the boundary ∂V together with any "external" body forces that may be acting. If we assume the fluid is inviscid, then the internal forces are just due to the pressure[3] which acts along the normal to ∂V. If there is a body force $\mathbf{F}$ per unit mass, and we suppose that we can apply Newton's equations to a volume of fluid, then

$$\frac{d}{dt}\left(\int_{V(t)} \rho \mathbf{u} \, dV\right) = -\int_{\partial V(t)} p\mathbf{n} \, dS + \int_{V(t)} \rho \mathbf{F} \, dV.$$

Using (2.3) on the left-hand side of this equation and the divergence theorem on the right-hand side, we obtain

$$\int_{V(t)} \left(\frac{d}{dt}(\rho \mathbf{u}) + \rho \mathbf{u}\nabla.\mathbf{u}\right) dV = \int_{V(t)} (-\nabla p + \rho \mathbf{F}) \, dV.$$

Remembering that this is true for any volume $V(t)$, and using (2.6), leads to

$$\frac{d\mathbf{u}}{dt} = \frac{\partial \mathbf{u}}{\partial t} + (\mathbf{u}.\nabla)\mathbf{u} = -\frac{1}{\rho}\nabla p + \mathbf{F}, \tag{2.7}$$

which is *Euler's equation* for an inviscid fluid.[4] If (2.6) and (2.7) both hold, it can be shown that the angular momentum of any volume V is also conserved (Exercise 2.3).

---

[3] It is at this stage that our restriction to inviscid flow is crucial. If the fluid has appreciable viscosity, the internal forces require much more careful consideration, as will be described briefly in Section 3.7, and more fully in Ockendon and Ockendon [41].

[4] Here we use $(\mathbf{u}.\nabla)\mathbf{u}$ to denote the operator $(\mathbf{u}.\nabla)$ in *Cartesian coordinates* acting on $\mathbf{u}$. In general coordinates, $(\mathbf{u}.\nabla)\mathbf{u}$ is $\frac{1}{2}\nabla|\mathbf{u}|^2 - \mathbf{u} \wedge (\nabla \wedge \mathbf{u})$.

For an incompressible fluid, (2.6) and (2.7) are sufficient to determine p and $\mathbf{u}$, but, when ρ varies, we need another relation involving p and ρ. This relation comes from considering conservation of energy, which will also involve the absolute temperature T, thus demanding yet another relation among p, ρ and T (the existence of an absolute zero of temperature will be discussed in the Appendix). When ρ is constant, the mechanical energy is automatically conserved if (2.6) and (2.7) are satisfied, and there is no need to consider energy conservation unless we are concerned with thermal effects.

In the simplest model for an inviscid compressible fluid, the energy consists of the *kinetic energy* of the fluid particles and the *internal energy* of the gas (potential energy will be accounted for separately if it is relevant). The internal energy represents the vibrational energy of the molecules of which the gas is composed and is manifested as the heat content of the gas. For an incompressible material, this heat content is the product of the specific heat and the absolute temperature, where the specific heat is determined from calorimetry. For a gas that can expand, we must take care that no unaccounted-for work is done by the pressure during the calorimetry, and so we insist that the experiment is done at constant volume. The resulting specific heat is denoted by c_v.

Now we must make a crucial assumption from thermodynamics. The *First Law of Thermodynamics* says that work, in the form of mechanical energy, can be transformed into heat, in the form of internal energy and vice versa, without any losses being incurred. Thus, we must add the internal and mechanical energies together so that the total local "energy density" is $e + \frac{1}{2}|\mathbf{u}|^2$, where $e = c_v T$ is the internal energy per unit mass. Now the rate of change of energy in a material volume V must be balanced against the following:

(i) The rate at which work is done by the body forces, and this is the term which will include the potential energy.
(ii) The rate at which work is done on the fluid volume by external forces.
(iii) The rate at which heat is transferred across ∂V.
(iv) The rate at which heat is created inside V by any source terms such as radiation.

By Fourier's Law, the rate at which heat is conducted in a direction $\mathbf{n}$ is $(-k\nabla T).\mathbf{n}$, where k is the conductivity of the material. Thus, conservation of energy for the fluid in $V(t)$ leads to the equation

$$\frac{d}{dt}\left[\int_{V(t)} (\frac{1}{2}\rho|\mathbf{u}|^2 + \rho e)\, dV\right]$$
$$= \int_V \rho\mathbf{F}.\mathbf{u}\, dV - \int_{\partial V} p\mathbf{u}.\mathbf{n}\, dS + \int_{\partial V} k\nabla T.\mathbf{n}\, dS$$
$$+ \frac{d}{dt}\left[\int_V \rho Q\, dV\right],$$

where Q is the heat addition per unit mass. Using the Transport Theorem (2.3), (2.6) and transforming the surface integrals by the divergence theorem, we obtain the equation

$$\rho \mathbf{u} . \frac{d\mathbf{u}}{dt} + \rho \frac{de}{dt} = -\nabla.(p\mathbf{u}) + \rho \mathbf{F}.\mathbf{u} + \nabla.(k\nabla T) + \rho \frac{dQ}{dt}. \tag{2.8}$$

This can be further simplified using (2.7) and (2.6) to get

$$\rho \frac{de}{dt} = \frac{p}{\rho} \frac{d\rho}{dt} + \nabla.(k\nabla T) + \rho \frac{dQ}{dt}, \tag{2.9}$$

(see Exercise 2.2).

We note that we can write the equations of conservation of mass and momentum, (2.6) and (2.7), in the form of a *conservation law*

$$\frac{\partial}{\partial t}(\text{density}) + \nabla.(\text{flux}) = (\text{external source}), \tag{2.10}$$

where the *density* and *flux* for mass conservation are ρ and $\rho \mathbf{u}$, respectively. For momentum conservation, the density is ρu_i (which we must, reluctantly, write using suffix notation) and the flux is the second-order tensor[5] $p\delta_{ij} + \rho u_i u_j$, where we define the divergence of the tensor a_{ij} as $\partial a_{ij}/\partial x_j$ and summation over the repeated suffix is assumed. Hence, we have

$$\text{div}(\rho u_i u_j + p\delta_{ij}) = \nabla.(\rho \mathbf{u})u_i + \rho(\mathbf{u}.\nabla)u_i + \frac{\partial p}{\partial x_i}, \tag{2.11}$$

which, on using (2.6), makes (2.10) equivalent to (2.7). The energy equation (2.8) can similarly be written in the form (2.10) with density $\rho(e + \frac{1}{2}|\mathbf{u}|^2)$ and the corresponding energy flux is

$$\rho \mathbf{u}(e + \frac{1}{2}|u|^2) + p\mathbf{u} - k\nabla T, \tag{2.12}$$

which is the sum of the convection of energy, the work done by pressure and the heat energy lost by conduction. The quantity $h = e + p/\rho$ is called the *enthalpy* of the gas, and the mechanical energy flux can alternatively be written as $\rho(h + \frac{1}{2}|\mathbf{u}|^2)\mathbf{u}$.

Looking back at (2.6), (2.7) and (2.9), we see that we have five formidable simultaneous nonlinear partial differential equations to solve. A first check shows that there are six dependent variables $\mathbf{u}, \rho, p$ and T and so, before we consider the appropriate boundary or initial conditions, we need to feed in some more information if we are to have any possibility of a well-posed mathematical model.

An immediate reaction is to note how much easier things are for an incompressible inviscid fluid. If we can say that ρ is constant, then the equations uncouple so that first (2.6) and (2.7) can be solved for p and $\mathbf{u}$ and (2.9) will determine T subsequently. Further than this, if we were considering a *barotropic* flow in

---

[5] Readers unfamiliar with tensor notation can refer, for instance, to Chapter 1 in [41]; δ_{ij} is the Kronecker delta.

which p is a prescribed function of ρ, then the same decomposition would occur.[6] Unfortunately most gas flows are far from barotropic, but there is one simple relationship that holds for gases that are not being compressed or expanded too violently. This is the *perfect gas law*

$$p = \rho RT. \tag{2.13}$$

It is both experimentally observed and predicted from statistical mechanics arguments that R is a universal constant.[7] The law applies to gases that are not so agitated that their molecules are out of thermodynamic equilibrium. If we assume the perfect gas law does hold, we are in effect requiring that any nonequilibrium effects are negligible, and we will discuss briefly how to model some nonequilibrium gas dynamics in Section 3.6. Furthermore, most observations to corroborate this law are made when the gas is at rest. This immediately raises the question of whether the relation (2.13) can be used to describe the gas dynamics we are modelling here, and in particular whether the pressure measured in static experiments can be identified with the variable p in (2.6), (2.7) and (2.9). For the moment, we will simply assume (2.13) is sufficient for practical purposes.

We are now almost in a position to make a dramatic simplification of (2.9). Before doing so, we need one other technical result that involves two "thought experiments". Suppose first that we change the state of a constant volume V of gas from pressure p and temperature T to pressure $p + \delta p$ and temperature $T + \delta T$. We assume that the gas is in equilibrium both at the beginning and end of this experiment. Then the amount of work needed to make this change is

$$\delta q = c_v \delta T. \tag{2.14}$$

Next we consider changing the state by altering V and T to $V + \delta V$ and $T + \delta T$ while keeping the pressure constant. In this case, the work needed to make this change is defined to be

$$\delta q' = c_p \delta T, \tag{2.15}$$

where c_p is the specific heat at constant pressure and, from (2.13),

$$p \delta V = R \delta T. \tag{2.16}$$

Finally we observe that, if we had attained this second state from the state $p + \delta p$, $T + \delta T$, V by an isothermal (constant temperature) change, we would have had to provide an extra amount of work $p \delta V$ over and above that needed for the constant volume change. Hence

$$\delta q' = \delta q + p \delta V$$

---

[6] Note that compressibility effects in water can be modelled by taking p proportional to ρ^γ where γ is about 7; see Glass and Sislan [22].

[7] It looks strange mathematically to put this constant in between two variables, but this is the conventional notation.

and so, from (2.14) and (2.15),

$$c_p \delta T = c_v \delta T + p \delta V.$$

Using (2.16), we find the relation

$$c_p - c_v = R. \tag{2.17}$$

It is conventional to define γ as the ratio of specific heats

$$\gamma = c_p/c_v \tag{2.18}$$

and we note that, since $R > 0$, $c_p > c_v$, and so $\gamma > 1$; it can be shown from the kinetic theory of gases that $\gamma = 1.4$ for nitrogen and this is approximately the value for air under everyday conditions.

For simplicity, let us assume that there is no heat conduction by putting $k = 0$ in (2.9) (this is part of the definition of an *ideal* gas). Then (2.9) becomes

$$\frac{de}{dt} - \frac{p}{\rho^2}\frac{d\rho}{dt} = \frac{dQ}{dt}, \tag{2.19}$$

and we can put $e = c_v T = c_v p/R\rho$ and $h = c_p T$, on using (2.13). Now the left-hand side of (2.19) depends only on p and ρ, and we can therefore find an integrating factor that makes this expression proportional to a total derivative. A simple calculation using (2.17) and (2.18) shows that

$$\begin{aligned}
\frac{de}{dt} - \frac{p}{\rho^2}\frac{d\rho}{dt} &= \frac{c_v}{R\rho}\frac{dp}{dt} - \left(\frac{c_v p}{R\rho^2} + \frac{p}{\rho^2}\right)\frac{d\rho}{dt} \\
&= \frac{c_v}{R\rho}\left[\frac{dp}{dt} - \frac{\gamma p}{\rho}\frac{d\rho}{dt}\right] \\
&= c_v T \frac{d}{dt}\left(\log\frac{p}{\rho^\gamma}\right).
\end{aligned}$$

Hence, if we write $S = S_0 + c_v \log(p/\rho^\gamma)$, where S_0 is a constant, we obtain the celebrated result

$$T\frac{dS}{dt} = \frac{dQ}{dt}. \tag{2.20}$$

The formal relation $T\delta S = \delta Q$ is the usual starting point for the definition of *entropy S* of a gas; when a unit mass of gas is heated by an amount δQ, its entropy is defined to be a function that changes by $\delta Q/T$. However, by starting from the energy equation we have shown that this mysterious function arises quite naturally in gas dynamics. The above discussion also enables us to state at once that, since radiative cooling has never been observed experimentally in a gas and since $T \geq 0$, then $dS/dt \geq 0$ which is a manifestation of the *Second Law of Thermodynamics*. It is not so easy to gain physical intuition about the entropy of a gas as it is, say, for temperature, but we will give a concrete illustration of its use in heat engines in the Appendix.

Finally, reinstating the conduction term in the energy equation, we can write (2.9) as

$$T\frac{dS}{dt} = \frac{1}{\rho}\nabla.(k\nabla T) + \frac{dQ}{dt}. \tag{2.21}$$

In most of the subsequent work, k and Q will be taken to be zero and so the equation will reduce to

$$\frac{dS}{dt} = 0. \tag{2.22}$$

In this situation, S is constant for a fluid particle and the flow is *isentropic*. If, in addition, the entropy of *all* fluid particles is the same (as would happen if the gas was initially uniform for instance), then $S \equiv S_0$ and the flow is *homentropic*.

In fact, the Second Law of Thermodynamics states that the *total* entropy of any thermodynamical system can never decrease, but here we have obtained the stronger statement (2.22) that the rate of change of entropy of any fluid particle is zero. Now it is well known (see, e.g., Ockendon & Ockendon [41]) that in any viscous flow in which there is shear, there is a positive dissipation of mechanical to thermal energy. Hence, we expect dS/dt to be positive whenever viscosity is present. On the other hand, as shown in Exercise 2.6, thermal conduction is a less powerful dissipative mechanism than viscosity because the equation $T(dS/dt) = (1/\rho)\nabla.(k\nabla T)$ does not constrain the sign of dS/dt.[8] We will return to these ideas in more detail in the Appendix.

We have now succeeded in writing down six equations, namely, (2.6), (2.7), (2.13) and (2.20), for our six dependent variables. Before considering their implications, we will consider briefly the sort of initial and boundary conditions that may arise.

2.2 Initial and Boundary Conditions

The presence of a single time derivative in each of (2.6), (2.7) and (2.9) suggests that, no matter what the boundary conditions are, we will require initial values for ρ, $\mathbf{u}$ and T and these will give the initial value for p from (2.13).

The boundary conditions are easy enough to guess when there is a prescribed impermeable boundary to the flow. We simply synthesise what is known about incompressible inviscid flow and heat conduction in solids to propose the following:

(i) The kinematic condition: the normal component of $\mathbf{u}$ should be equal to the normal velocity of the boundary (with no condition on p).
(ii) The thermodynamic condition: the temperature or the heat flux, $-k\mathbf{n}.\nabla T$, or some combination of these two quantities should be prescribed. This assumes that $k > 0$; if $k = 0$, then no thermodynamic condition is needed.

---

[8] We hasten to emphasise that in most gases, the effects of viscosity and thermal conductivity are of comparable size: hence, the study of an inviscid gas with $k > 0$ is of purely academic interest.

For a prescribed, moving, impermeable boundary $f(\mathbf{x}, t) = 0$, we note that a consequence of the continuum hypothesis is that fluid particles which are on the boundary of a fluid at any time must always remain on the boundary. Hence, the kinematic condition on the boundary is

$$\frac{df}{dt} = 0 = \frac{\partial f}{\partial t} + \mathbf{u}.\nabla f. \tag{2.23}$$

However, the situation becomes much more complicated when the boundary of the gas is free rather than being prescribed. This could occur if the gas was flowing through a shock wave and this difficult situation will be discussed in Chapter 6. Things are simpler for an incompressible flow, such as the flow of water with a free surface; now we must impose a second condition over and above the kinematic condition (2.23) if we are to be able to solve the field equations and *also* determine the position of the boundary. This second condition comes from considering the momentum balance. A simple argument suggests that, in the absence of surface tension, the pressure must be continuous across the boundary, because the boundary has no inertia; hence,

$$p_1 = p_2 \tag{2.24}$$

on the boundary, where p_2 is the external pressure and p_1 is the pressure in the fluid. Conditions (2.23) and (2.24) will be reconsidered more carefully in specific circumstances in later chapters.

Before considering the full implications of the model we have derived, it is very helpful to recall some well-known results about incompressible flow. This will not only help us pose the best questions to ask about compressible flows in general but will also provide useful background for some of the models to be considered in Chapter 3.

2.3 Vorticity and Irrotationality

2.3.1 Homentropic Flow

One distinctive attribute of fluid mechanics, compressible or incompressible, compared to other branches of continuum mechanics is the existence of *vorticity* ω, defined by $\omega = \nabla \wedge \mathbf{u}$. We can derive an equation for the evolution of ω by first writing

$$(\mathbf{u}.\nabla)\mathbf{u} = \frac{1}{2}\nabla|\mathbf{u}|^2 - \mathbf{u} \wedge (\nabla \wedge \mathbf{u})$$

in (2.7). If we assume that $\mathbf{F}$ is a conservative force so that $\mathbf{F} = -\nabla\Omega$ for some scalar potential Ω, and we use the same algebraic manipulations as those used to derive (2.20), we obtain

$$\frac{d\mathbf{u}}{dt} = \frac{\partial \mathbf{u}}{\partial t} + \nabla(\frac{1}{2}|\mathbf{u}|^2) - \mathbf{u} \wedge \omega = \nabla(-\Omega - \frac{\gamma p}{(\gamma - 1)\rho}) + T\nabla S. \tag{2.25}$$

Taking the curl of this equation leads to

$$\frac{\partial \boldsymbol{\omega}}{\partial t} + (\mathbf{u}.\nabla)\boldsymbol{\omega} + (\nabla.\mathbf{u})\boldsymbol{\omega} = (\boldsymbol{\omega}.\nabla)\mathbf{u} + \nabla \wedge (T\nabla S).$$

Hence, using (2.6),

$$\frac{\partial}{\partial t}\left(\frac{\boldsymbol{\omega}}{\rho}\right) + (\mathbf{u}.\nabla)\frac{\boldsymbol{\omega}}{\rho} = \frac{1}{\rho}(\boldsymbol{\omega}.\nabla)\mathbf{u} + \frac{1}{\rho}\nabla T \wedge \nabla S. \tag{2.26}$$

For a homentropic flow, ∇S will be zero and so

$$\frac{d}{dt}\left(\frac{\boldsymbol{\omega}}{\rho}\right) = \left(\frac{\boldsymbol{\omega}}{\rho}.\nabla\right)\mathbf{u}. \tag{2.27}$$

Thus, in two-dimensional homentropic flow, in which $(\boldsymbol{\omega}.\nabla)\mathbf{u}$ is automatically zero, $\boldsymbol{\omega}/\rho$ is convected with the fluid. Remarkably, if we change to Lagrangian variables, (2.27) can be solved explicitly, even in three dimensions (see Exercise 2.4), to give

$$\frac{\boldsymbol{\omega}}{\rho} = \left(\frac{\boldsymbol{\omega}_0}{\rho_0}.\nabla_{\mathbf{a}}\right)\mathbf{x}, \tag{2.28}$$

where $\nabla_{\mathbf{a}}$ is the gradient operator with respect to Lagrangian variables $\mathbf{a}$ and $\boldsymbol{\omega}_0$, ρ_0 are the values of $\boldsymbol{\omega}, \rho$ at $t = 0$. This is *Cauchy's equation* for the vorticity in an arbitrary homentropic flow, but it is not very useful since we cannot find $\nabla_{\mathbf{a}}$ until we have found the flow field! However, (2.28) does tell us immediately that if the vorticity is everywhere zero in a fluid region $V(0)$ at $t = 0$, then it will be zero at all subsequent times in the region $V(t)$ which contains the same fluid particles as $V(0)$. Thus, $\boldsymbol{\omega} \equiv \mathbf{0}$ in $V(t)$ and the flow is *irrotational*. Such flows occur, for example, when the fluid is initially at rest or when there are uniform conditions at infinity in steady flow.

To understand vorticity transport geometrically, we assume ρ is constant for simplicity and plot the trajectories of two nearby fluid particles that are at $\mathbf{x}(t)$ and $\mathbf{x}(t) + \varepsilon \boldsymbol{\omega}(\mathbf{x}(t), t)$ at time t, as shown in Figure 2.1. After a short time δt, the particles will have moved to $\mathbf{x}(t) + \mathbf{u}(\mathbf{x}, t)\delta t$ and $\mathbf{x}(t) + \varepsilon \boldsymbol{\omega}(\mathbf{x}, t) + \mathbf{u}(\mathbf{x} + \varepsilon \boldsymbol{\omega}(\mathbf{x}, t), t)\delta t$, respectively, and the vector joining the two particles will therefore have changed from $\varepsilon \boldsymbol{\omega}(\mathbf{x}, t)$ to $\varepsilon \boldsymbol{\omega}(\mathbf{x}, t) + \varepsilon(\boldsymbol{\omega}.\nabla)\mathbf{u}\delta t$. However, from (2.27) with ρ constant,

$$(\boldsymbol{\omega}.\nabla)\mathbf{u}\delta t = \boldsymbol{\omega}(\mathbf{x}(t + \delta t), t + \delta t) - \boldsymbol{\omega}(\mathbf{x}(t), t),$$

and so the vector joining the particles at $t + \delta t$ is $\varepsilon \boldsymbol{\omega}(\mathbf{x}(t + \delta t), t + \delta t)$. Thus, we can see that, in three dimensions, the *vortex lines*, which are parallel to the vorticity at each point of the fluid, move with the fluid and are stretched as the vorticity increases.

An alternative way to approach vorticity is to consider the total vorticity flux through an arbitrary closed contour $C(t)$ which moves with the fluid. This quantity, known as the *circulation* round C, is given by

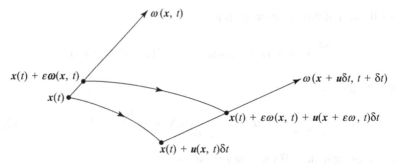

Fig. 2.1 Convection of vorticity in an incompressible fluid

$$\Gamma = \int_C \mathbf{u}.d\mathbf{x} = \int_\Sigma \boldsymbol{\omega}.d\mathbf{S},$$

where Σ is any smooth surface spanning C and contained within the fluid. Note that the circulation integral round C is defined even in a non-simply connected region. To consider the rate of change of Γ, we change to Lagrangian variables so that

$$\Gamma = \int_{C(t)} u_i \, dx_i = \int_{C(0)} u_i \frac{\partial x_i}{\partial a_j} da_j.$$

Then

$$\frac{d\Gamma}{dt} = \int_{C(0)} \frac{du_i}{dt} \frac{\partial x_i}{\partial a_j} da_j + u_i \frac{\partial \dot{x}_i}{\partial a_j} da_j$$

$$= \int_{C(t)} \left(\frac{d\mathbf{u}}{dt}.d\mathbf{x} + \mathbf{u}.d\mathbf{u} \right).$$

Now

$$\int_{C(t)} \mathbf{u}.d\mathbf{u} = \left[\frac{1}{2}(\mathbf{u})^2 \right]_{C(t)} = 0$$

since $\mathbf{u}$ is a single-valued function and so, using (2.25),

$$\frac{d\Gamma}{dt} = \int_{C(t)} T\nabla S.d\mathbf{x} - \left[\Omega + \frac{\gamma p}{(\gamma - 1)\rho} \right]_{C(t)}$$

$$= \int_{C(t)} T\nabla S.d\mathbf{x},$$

since Ω, p and ρ are all single-valued functions. For a homentropic flow, $\nabla S = \mathbf{0}$ and we have *Kelvin's Theorem* which shows that the circulation round any closed contour moving with the fluid is constant. In particular, if the fluid region is simply connected, we again arrive at the result that if $\boldsymbol{\omega} \equiv \mathbf{0}$ at $t = 0$ for all points in a simply connected region, then $\Gamma \equiv 0$ for all closed curves C and so the flow is irrotational.[9]

Note that we can use the identity $\nabla \wedge (T\nabla S) = \nabla T \wedge \nabla S$ to write Kelvin's Theorem in the form

$$\frac{d\Gamma}{dt} = \int_{\Sigma} (\nabla T \wedge \nabla S).d\mathbf{S}.$$

Now, in any smooth irrotational flow in a simply connected region, Γ is identically zero and so $\nabla T \wedge \nabla S = \mathbf{0}$. Since T is proportional to p/ρ and S is a function of p/ρ^γ, with $\gamma > 1$, T cannot be a function of S alone and so the flow must be either homentropic or isothermal. The latter is unlikely in practice, and vorticity can thus be associated with an entropy gradient and vice versa except in special cases (see Exercise 2.5).

Whenever the flow is irrotational, we can define a *velocity potential* ϕ by $\phi(\mathbf{x}, t) = \int_{\mathbf{x}_0}^{\mathbf{x}} \mathbf{u}.d\mathbf{x}$ for any convenient constant $\mathbf{x}_0$, and, from Kelvin's theorem, ϕ will be a well-defined function of $\mathbf{x}$ and t. From this definition we can write

$$\mathbf{u} = \nabla\phi.$$

Now substituting for $\mathbf{u}$ in (2.6) and (2.25), the equations for homentropic irrotational flow with a conservative body force collapse to

$$\frac{\partial\rho}{\partial t} + \nabla.(\rho\nabla\phi) = 0 \qquad (2.29)$$

and

$$\frac{\partial\phi}{\partial t} + \frac{1}{2}|\nabla\phi|^2 + \Omega + \frac{\gamma p}{(\gamma - 1)\rho} = G(t), \qquad (2.30)$$

where G is some function of t, often determined by the conditions at infinity. Equation (2.30) is *Bernoulli's equation* for homentropic gas flow.

2.3.2 Incompressible Flow

Most of the modelling in the previous section is an obvious generalisation of well-known results for inviscid incompressible flows. In particular, homentropic compressible flow has many features in common with incompressible flow; (2.27) and (2.28) hold for incompressible flow as does Kelvin's theorem, and in both cases the existence of a velocity potential in irrotational flow leads to a dramatic simplification.

However, the incompressible limit of our compressible model is nontrivial mathematically and we only make one general remark about it here, though we will

---

[9] It is easy to see that this result does not apply in, say, a circular annulus when $\mathbf{u} = (\Gamma/2\pi)\mathbf{e}_\theta$ in polar coordinates.

return to it again in Section 3.1. We have already noted in footnote 6 on p. 11 that one possible procedure is to let $\gamma \to \infty$. Now γ only enters the general model via the energy equation (2.22) which we can write as

$$\frac{d}{dt}\left(\frac{\rho}{p^{1/\gamma}}\right) = 0.$$

Letting $\gamma \to \infty$ now clearly suggests that $d\rho/dt = 0$ and hence that the flow is incompressible. We also note that letting $\gamma \to \infty$ in (2.30) leads to the familiar incompressible form of Bernoulli's equation.

In the next chapter, we will use our *nonlinear* model for gas dynamics as a basis for the *linearised* theory of *acoustics* or *sound waves*. This will lead us to the prototype of all models for wave motion. Even more importantly, it will show how the linearisation of an intractable nonlinear problem can lead to a linear wave propagation model which is both revealing and relatively straightforward to analyse.

Exercises

R 2.1 If J is the Jacobian $\partial(x_1, x_2, x_3)/\partial(a_1, a_2, a_3)$, where $\mathbf{a}$ are Lagrangian coordinates, use (2.2) and (2.6) to show that $d(\rho J)/dt = 0$.

Hence, show that the equations of one-dimensional gas dynamics in Lagrangian coordinates are

$$\rho \frac{\partial x}{\partial a} = \rho_0 \quad \text{and} \quad \rho_0 \frac{\partial u}{\partial t} + \frac{\partial p}{\partial a} = 0,$$

where $u = \partial x/\partial t$.

R 2.2 The equations for a compressible gas, written in conservation form, are, in the absence of heat conduction or radiation,

$$\frac{\partial \rho}{\partial t} + \nabla.(\rho \mathbf{u}) = 0, \tag{1}$$

$$\frac{\partial}{\partial t}(\rho \mathbf{u}) + (\mathbf{u}.\nabla)(\rho \mathbf{u}) + \rho \mathbf{u} \nabla.\mathbf{u} = -\nabla p, \tag{2}$$

and

$$\frac{\partial}{\partial t}(\rho(e + \frac{1}{2}|\mathbf{u}|^2)) + \nabla.(\rho(e + \frac{1}{2}|\mathbf{u}|^2)\mathbf{u}) = -\nabla.(p\mathbf{u}). \tag{3}$$

From (1) and (2) show that the Euler equation

$$\frac{d\mathbf{u}}{dt} = -\frac{1}{\rho}\nabla p \tag{4}$$

holds. Using (1) and (4) to eliminate $d\rho/dt$ and $d\mathbf{u}/dt$ from (3), show that

$$\rho \frac{de}{dt} = -p\nabla.\mathbf{u},$$

and hence, from (1), that

$$\frac{de}{dt} = \frac{p}{\rho^2}\frac{d\rho}{dt}.$$

Deduce that p/ρ^γ is a constant for a fluid particle in a perfect gas.

2.3 Define the angular momentum of a material volume V as

$$\mathbf{L} = \int_{V(t)} \mathbf{x} \wedge \rho \mathbf{u} \, dV,$$

where $\mathbf{x}$ is the position of a particle of fluid with respect to a fixed origin. Show by using equations (2.6) and (2.7) that

$$\frac{d\mathbf{L}}{dt} = -\int_{\partial V(t)} \mathbf{x} \wedge p\mathbf{n} \, dS + \int_{V(t)} \mathbf{x} \wedge \rho \mathbf{F} \, dV,$$

and deduce that the rate of change of angular momentum of the fluid in $V(t)$ is equal to the sum of the moments of the forces acting on $V(t)$.

Note that if this formula is applied to the angular momentum of a small element of fluid Σ about its centre of gravity, the magnitude of $\mathbf{L}$ will be of $O(\delta^4)$ if δ is the length scale of the element, whereas the term $\int \mathbf{x} \wedge p\mathbf{n} dS$ is of $O(\delta^3)$. Letting $\delta \to 0$, this gives

$$p \int_{\partial \Sigma} \mathbf{x} \wedge \mathbf{n} \, dS = \mathbf{0},$$

which, fortunately, is identically true.

2.4 Starting from the Euler equation (2.7) with $\mathbf{F} = \mathbf{0}$, show that, in homentropic flow, $\tilde{\boldsymbol{\omega}} = \nabla \wedge \mathbf{u}/\rho$ satisfies the equation

$$\frac{d\tilde{\boldsymbol{\omega}}}{dt} = (\tilde{\boldsymbol{\omega}}.\nabla)\mathbf{u}.$$

By changing to Lagrangian variables $\mathbf{a}, t$, where $\mathbf{x}(\mathbf{a}, 0) = \mathbf{a}$, show that

$$\frac{d\tilde{\omega}_i}{dt} = \tilde{\omega}_k \frac{\partial a_j}{\partial x_k} \frac{d}{dt}\left(\frac{\partial x_i}{\partial a_j}\right),$$

where the summation convention for the repeated suffices j and k is used. Noting that $(\partial x_i/\partial a_k).(\partial a_k/\partial x_j) = \delta_{ij}$, show that

$$\frac{d}{dt}\left(\tilde{\omega}_k \frac{\partial a_i}{\partial x_k}\right) = 0,$$

and hence deduce that

$$\tilde{\boldsymbol{\omega}} = (\tilde{\boldsymbol{\omega}}_0.\nabla_a)\mathbf{x},$$

where $\tilde{\boldsymbol{\omega}} = \tilde{\boldsymbol{\omega}}_0$ at $t = 0$.

R 2.5 Show that in a *two-dimensional* steady flow, the entropy S is constant on a streamline and hence that $\mathbf{u}$ and $\nabla \wedge \mathbf{u}$ are perpendicular to ∇S. Deduce *Crocco's Theorem* which states that, for rotational, non-homentropic flow,

$$\mathbf{u} \wedge (\nabla \wedge \mathbf{u}) = \lambda \nabla S$$

for some scalar function λ.

R 2.6 Show that in a heat conducting gas with positive conductivity k (which need not be constant)

$$T\frac{dS}{dt} = \frac{1}{\rho}\nabla.(k\nabla T).$$

Deduce that if the gas is confined in a fixed thermally insulated container Ω, then the rate of change of total entropy is

$$\frac{d}{dt}\left[\int_{\Omega} \rho S \, dV\right] = \int_{\Omega} \frac{k|\nabla T|^2}{T^2} \, dV \geq 0.$$

2.7 If Ω is an arbitrary volume of fluid *fixed in space*, show that the principle of conservation of mass implies that

$$\frac{d}{dt}\int_{\Omega} \rho \, dV = -\int_{\partial \Omega} \rho \mathbf{u}.d\mathbf{S}$$

and hence deduce (2.6). In a similar way, deduce (2.7) and (2.9) by considering the momentum and energy of the fluid in Ω.

Chapter 3
Models for Linear Wave Propagation

This chapter will discuss models for several quite different classes of waves with the common characteristic that the models can be linearised for waves of sufficiently small amplitude. We will focus on waves in fluids, but even here we will find that the models are far from trivial and can look very different from each other. Their unifying features will become more apparent when we embark on their mathematical analysis in Chapter 4. We begin with sound waves which are one of the most familiar of all waves.

3.1 Acoustics

The theory of acoustics is based on the fact that in sound waves (at least those that do not affect the eardrum adversely), the variations in pressure, density, temperature are all small compared to some ambient conditions. These ambient conditions from which the motion is initiated are usually either that the gas is at rest, so that $p = p_0$, $\rho = \rho_0$, $T = T_0$ and $\mathbf{u} = \mathbf{0}$, or the gas is in a state of uniform motion in which $\mathbf{u} = U\mathbf{i}$, say. We start with the simplest case and motivate the linearisation procedure in an intuitive way.

We suppose that the gas is initially at rest in a long pipe along the x-axis and that it is subject to a small disturbance so that

$$\mathbf{u} = \bar{u}(x, t)\mathbf{i}.$$

We assume that $\bar{p} = p - p_0$ and $\bar{\rho} = \rho - \rho_0$ are all "small" and neglect the squares of the barred quantities. From equations (2.6) and (2.7), we find

$$\frac{\partial \bar{\rho}}{\partial t} + \rho_0 \frac{\partial \bar{u}}{\partial x} = 0 \qquad (3.1)$$

© Springer Science+Business Media New York 2015
H. Ockendon, J.R. Ockendon, *Waves and Compressible Flow*,
Texts in Applied Mathematics 47, DOI 10.1007/978-1-4939-3381-5_3

and

$$\frac{\partial \bar{u}}{\partial t} + \frac{1}{\rho_0} \frac{\partial \bar{p}}{\partial x} = 0. \tag{3.2}$$

The energy equation (2.22) reduces to $p/\rho^\gamma = p_0/\rho_0^\gamma$, so that

$$\bar{p} = \frac{\gamma p_0}{\rho_0} \bar{\rho} \tag{3.3}$$

to a first approximation. We define c_0^2 to be $\gamma p_0/\rho_0$, and then, from (3.1), (3.2) and (3.3), we can show that the variables $\bar{\rho}$, $\bar{u}$, $\bar{p}$ all satisfy the same equation, namely,

$$\frac{\partial^2 \phi}{\partial x^2} = \frac{1}{c_0^2} \frac{\partial^2 \phi}{\partial t^2}; \tag{3.4}$$

this is the well-known one-dimensional wave equation which generates waves travelling with speed $\pm c_0$ and c_0 is known as the *speed of sound*.

The simplicity of (3.4) in comparison with (2.6), (2.7) and (2.9) is dramatic and the validity of the linearisation procedure requires careful scrutiny. In fact, even assuming we are in a regime where (2.6), (2.7) and (2.9) are valid, much more care is needed to derive (3.4) than the simple assumption that the square of the perturbations (the barred variables) can be neglected. Most strikingly, even though $\bar{u}$ is small, $\partial \bar{u}/\partial x$ may be large so that the neglect of the nonlinear term $\bar{u}(\partial \bar{u}/\partial x)$ may not be justified. Also, not only must the amplitude of the waves be small, but the time variation must not be too slow if it is to interact with the spatial variation. In order to clarify the assumptions built into the approximation represented by (3.4), we need to do a systematic nondimensionalisation and analyse the equations as below.

In many circumstances, the wave motion will be driven with a prescribed velocity u_0, frequency ω_0 and propagate over a known length scale L. We therefore introduce nondimensional variables

$$\rho = \rho_0(1 + \varepsilon \hat{\rho}),$$
$$p = p_0(1 + \varepsilon \hat{p}),$$
$$u = u_0 \hat{u},$$
$$x = LX$$

and

$$t = \omega_0^{-1} T,$$

where ε is a small dimensionless parameter. Then (2.6) and (2.7) become

$$\varepsilon \rho_0 \omega_0 \frac{\partial \hat{\rho}}{\partial T} + \frac{\rho_0 u_0}{L} \frac{\partial \hat{u}}{\partial X} + \frac{\rho_0 u_0 \varepsilon}{L} \frac{\partial}{\partial X}(\hat{u}\hat{\rho}) = 0$$

and

$$(1 + \varepsilon \hat{\rho}) \left(u_0 \omega_0 \frac{\partial \hat{u}}{\partial T} + \frac{u_0^2}{L} \hat{u} \frac{\partial \hat{u}}{\partial X} \right) = -\frac{\varepsilon p_0}{L \rho_0} \frac{\partial \hat{p}}{\partial X}.$$

These equations will thus retain the same terms as (3.1) and (3.2), as a first approximation in ε, if[1] $u_0 \sim \varepsilon \omega_0 L \sim \varepsilon p_0 / L \omega_0 \rho_0$ and, remembering that $c_0^2 = \gamma p_0 / \rho_0$, this is achieved by taking

$$\omega_0 L \sim c_0, \quad u_0 \sim \varepsilon c_0. \tag{3.5}$$

Thus, for example, if the motion is being driven by a piston oscillating with speed u_0, then u_0 must be much smaller than the speed of sound in the undisturbed gas for the linearisation to be valid. If ε is defined to be u_0 / c_0, then the resulting pressure and density variations will automatically be of $O(\varepsilon p_0)$ and $O(\varepsilon \rho_0)$. Equally, if the motion is driven by a prescribed pressure oscillation of amplitude $O(\varepsilon p_0)$, then the resulting density and velocity changes will be $O(\varepsilon \rho_0)$ and $O(\varepsilon c_0)$. In all cases, our theory will only describe waves whose frequency is no higher than $O(c_0 / L)$.

Although this derivation of (3.4) is more laborious than the simple handwaving that we used at the beginning of the section, it is the only way we can have any reliable knowledge of the range of validity of the model and we will need to take this degree of care throughout this chapter.

We note here some other important but less fundamental remarks about the acoustic approximation.

(i) **Sound waves in three dimensions.**

As shown in Exercise 3.1, in higher dimensions (3.4) is replaced by[2]

$$\nabla^2 \phi = \frac{1}{c_0^2} \frac{\partial^2 \phi}{\partial t^2}. \tag{3.6}$$

This may still reduce to a problem in two variables if we have either circular symmetry, when $\nabla^2 = \partial^2 / \partial r^2 + (1/r)\partial/\partial r$, or spherical symmetry, when $\nabla^2 = \partial^2 / \partial r^2 + (2/r)\partial/\partial r$, in suitable polar coordinates.

(ii) **Sound waves in a medium moving with uniform speed U.**

If the uniform flow U is taken along the x axis, it can be shown (Exercise 3.1) that, by writing $\mathbf{u} = U\mathbf{i} + \varepsilon \nabla \phi$, (3.4) is now replaced by

$$\nabla^2 \phi = \frac{1}{c_0^2} \left(\frac{\partial}{\partial t} + U \frac{\partial}{\partial x} \right)^2 \phi.$$

In particular, for steady flow,

$$\left(1 - \frac{U^2}{c_0^2} \right) \frac{\partial^2 \phi}{\partial x^2} + \frac{\partial^2 \phi}{\partial y^2} + \frac{\partial^2 \phi}{\partial z^2} = 0, \tag{3.7}$$

---

[1] Here we use the symbol $\sim$ to mean "is of roughly the same size as" or, more precisely, "is of the order of" as $\varepsilon \to 0$.

[2] Note that the three-dimensional version of (3.2) is $\partial \mathbf{u}/\partial t = -(1/\rho_0)\nabla p$, which automatically guarantees that $\partial \boldsymbol{\omega}/\partial t = \mathbf{0}$; this makes irrotationality even more common than Kelvin's theorem suggests.

and it is clear that the parameter U/c_0 now plays a key role in the solution. It is called the *Mach number* and the flow is *supersonic* if $M > 1$ ($U > c_0$) and *subsonic* if $M < 1$ ($U < c_0$). Note that the Mach number of acoustic waves in a stationary medium is of $O(\varepsilon)$ by (3.5), even though the waves themselves propagate at sonic speed.

3.2 Surface Gravity Waves in Incompressible Flow

We now consider the problem of waves on the surface of an incompressible fluid subject to gravitational forces. It may seem strange to suddenly revert to incompressible flow at this stage, but, in fact, we can think of water and air separated by an interface as an extreme case of a variable density fluid where all the density variation takes place at the surface. The ratio of densities of air and water is about 10^{-3}, so the jump is extreme in magnitude as well as occurring over a very short distance. We will come back to this point of view later, but, for the moment, we will derive the governing equations from the usual equations of incompressible fluid dynamics.

We recalled in Chapter 2 that the classical theory of inviscid flow predicts that if the fluid motion is initially irrotational, then it will remain irrotational. Thus, writing $\mathbf{u} = \nabla\phi$, the field equations reduce to Laplace's equation

$$\nabla^2\phi = 0 \tag{3.8}$$

for ϕ and to Bernoulli's equation

$$\frac{\partial\phi}{\partial t} + \frac{1}{2}|\nabla\phi|^2 + gz + \frac{p}{\rho} = \frac{p_0}{\rho} \tag{3.9}$$

for p, where we have assumed that the external pressure in the air is p_0 and that the z-axis is vertical. What is important now are the boundary conditions for ϕ at the free surface. We anticipate that, whereas only one condition is needed for ϕ at a prescribed boundary, we will now need two conditions to compensate for the fact that the position of the free surface is unknown and needs to be determined as part of the solution of the problem. A problem of this type is known as a *free boundary problem*.

The first free surface condition comes from the fact that no fluid particle can cross the surface (we will neglect any "spray"). If the surface is given by $z = \eta(x, t)$, where we are considering a two-dimensional situation for simplicity, a particle on the surface has position $(x, 0, \eta)$ and the velocity of this particle is $(u, 0, w)$, where

$$w = \frac{d\eta}{dt} = \frac{\partial\eta}{\partial t} + u\frac{\partial\eta}{\partial x}.$$

Hence, as could have also been deduced from (2.23), we have the *kinematic* boundary condition

$$\frac{\partial \phi}{\partial z} = \frac{\partial \eta}{\partial t} + \frac{\partial \phi}{\partial x}\frac{\partial \eta}{\partial x}, \tag{3.10}$$

which expresses the principle of conservation of mass at the free surface.

The second condition expresses the principle of conservation of momentum at the free surface. As discussed in Chapter 2, this simply means that if surface tension effects can be neglected,[3] then the pressure at the surface will be p_0 so that from (3.9)

$$\frac{\partial \phi}{\partial t} + \frac{1}{2}|\nabla \phi|^2 + g\eta = 0 \tag{3.11}$$

on $z = \eta(x, t)$.

If we apply suitable initial conditions (which must satisfy irrotationality), and conditions at any fixed boundaries, we will have a fully nonlinear model for surface gravity waves. This model is every bit as formidable as the compressible equations (2.6), (2.7) and (2.22), so let us again consider the effect of linearisation. We will take water of depth h at rest as the basic equilibrium state and formally neglect squares and products of the variables ϕ and η. There is one extra subtlety here because when we make this assumption in (3.10), we must, to be consistent, write

$$\frac{\partial \phi}{\partial z} = \frac{\partial \eta}{\partial t} \quad \text{on } z = 0,$$

rather than on $z = \eta$. This is because the difference between $\partial \phi/\partial z(x, \eta, t)$ and $\partial \phi/\partial z(x, 0, t)$ is a product of η and $\partial^2 \phi/\partial z^2$ and thus is negligible under the linearisation approximation. Hence, from (3.8), (3.10) and (3.11), the formal model for small-amplitude waves, called *Stokes waves*, on water of depth h is

$$\nabla^2 \phi = 0, \tag{3.12}$$

with

$$\frac{\partial \phi}{\partial z} = \frac{\partial \eta}{\partial t}, \quad \frac{\partial \phi}{\partial t} + g\eta = 0 \quad \text{on } z = 0 \tag{3.13}$$

and

$$\frac{\partial \phi}{\partial z} = 0 \quad \text{on } z = -h. \tag{3.14}$$

The conditions (3.13) can be further reduced to a single condition on ϕ in the form

$$\frac{\partial^2 \phi}{\partial t^2} + g\frac{\partial \phi}{\partial z} = 0 \quad \text{on } z = 0, \tag{3.15}$$

and we are left with the problem of solving Laplace's equation (3.8) with an odd-looking boundary condition (3.15) on one prescribed boundary and a more standard condition (3.14) on the other. Although linearisation has greatly simplified the

---

[3] See Exercise 4.6 for a brief discussion of the effect of surface tension.

difficulty caused by the free boundary, (3.15) poses a new challenge. Standard theory tells us that Laplace's equation can usually be solved uniquely, or to within a constant, if ϕ or its normal derivative or even a linear combination thereof is prescribed on the boundary of a closed region, but (3.15) does not fall into any of these categories.

Before making any further remarks about this model, we will repeat the procedure adopted in Section 3.1 for discussing the parameter regime in which we might expect (3.12)–(3.15) to be valid. We suppose that the disturbance to the surface of the water has an amplitude a, which must be small compared to the depth h. Then, we nondimensionalise by introducing an arbitrary length scale λ, time scale ω_0^{-1} and potential scale ϕ_0 and writing $\eta = a\hat{\eta}$, $x = \lambda X$, $z = \lambda Z$, $t = \omega_0^{-1}T$ and $\phi = \phi_0\hat{\phi}$. We find that the linearised equations (3.12)–(3.15) are a valid approximation provided λ, ω_0 and ϕ_0 satisfy

$$\omega_0 = \left(\frac{g}{\lambda}\right)^{1/2}, \quad \phi_0 = a(\lambda g)^{1/2} \quad \text{and} \quad \frac{a}{\lambda} \ll 1. \tag{3.16}$$

Since the boundary condition (3.14) is applied on $Z = -h/\lambda$, we will also need to insist that $h/\lambda \geq O(1)$. If this latter restriction is violated, we can still make simplifications, and these lead to the nonlinear *shallow water* theory as will be described in Chapter 5.

Once again, we can extend this theory easily enough to three dimensions when equations (3.12)–(3.15) will still be valid as long as we write $\nabla^2\phi$ as $\partial^2\phi/\partial x^2 + \partial^2\phi/\partial y^2 + \partial^2\phi/\partial z^2$. It is also straightforward to consider waves on a uniform stream moving with velocity $U\mathbf{i}$ and in this case the only change is that (3.15) becomes

$$\left(\frac{\partial}{\partial t} + U\frac{\partial}{\partial x}\right)^2 \phi + g\frac{\partial\phi}{\partial z} = 0.$$

3.3 Internal Waves

As a generalisation of the last section, we now consider flows which consist of incompressible particles, but where the density may vary from particle to particle. This may arise, for example, in oceanography, where the density of the sea is related to the salinity and diffusion is so small that the salinity of a fluid particle is conserved. Thus,

$$\frac{d\rho}{dt} = \frac{\partial\rho}{\partial t} + \mathbf{u}.\nabla\rho = 0, \tag{3.17}$$

and (2.6) and (2.7) reduce to

$$\nabla.\mathbf{u} = 0 \tag{3.18}$$

and

$$\frac{\partial\mathbf{u}}{\partial t} + (\mathbf{u}.\nabla)\mathbf{u} = -\frac{1}{\rho}\nabla p - g\mathbf{k}, \tag{3.19}$$

where **k** is measured vertically upwards. We now have sufficient equations to solve for **u**, p and ρ. Moreover, using (2.7) in the energy equation removes the terms involving the gravitational body force and reduces (2.9) to

$$\frac{de}{dt} = 0.$$

Thus, when there is no conduction, the temperature is constant for each fluid particle.

An exact hydrostatic solution of equations (3.17), (3.18) and (3.19) is that of a *stratified* fluid where

$$\mathbf{u} = 0, \quad \rho = \rho_s(z) \quad \text{and} \quad p = p_0 - g \int_0^z \rho_s(\sigma)\, d\sigma = p_s(z), \qquad (3.20)$$

say, where p_0 is a constant reference pressure on $z = 0$. Now we can, as usual, effect a handwaving derivation of the linear theory about the state given by (3.20). For simplicity, we look at two-dimensional disturbances and assume that $\bar{\rho} = \rho - \rho_s(z)$, $\bar{p} = p - p_s(z)$ and $|\mathbf{u}| = |(u, 0, w)|$ are all small. Then, with $' = d/dz$, (3.17), (3.18) and (3.19) reduce to

$$\frac{\partial \bar{\rho}}{\partial t} + \rho_s' w = 0, \qquad (3.21)$$

$$\frac{\partial u}{\partial x} + \frac{\partial w}{\partial z} = 0, \qquad (3.22)$$

$$\rho_s \frac{\partial u}{\partial t} = -\frac{\partial \bar{p}}{\partial x} \qquad (3.23)$$

and

$$\rho_s \frac{\partial w}{\partial t} = -\frac{\partial \bar{p}}{\partial z} - \bar{\rho} g. \qquad (3.24)$$

It is now a simple matter to cross-differentiate to eliminate $\bar{\rho}$, $\bar{p}$ and u to obtain

$$\frac{\partial^2}{\partial t^2}\left(\frac{\partial^2 w}{\partial x^2} + \frac{\partial^2 w}{\partial z^2}\right) = -N^2(z)\left(\frac{\partial^2 w}{\partial x^2} - g^{-1}\frac{\partial^3 w}{\partial z \partial t^2}\right), \qquad (3.25)$$

where $N^2(z) = -g(\rho_s(z)'/\rho_s(z))$ is a positive function in a stably stratified fluid. We note with satisfaction that if

$$\rho_s(z) = \begin{cases} 0, & z > 0 \\ \rho_0, & z < 0 \end{cases},$$

as was the case in Section 3.2, then, in $z < 0$, w will be a potential function (assuming suitable initial conditions). Moreover, by integrating (3.22) across $z = 0$, we find that w is continuous there and from (3.24) we get that $\bar{p}$ is also continuous, which are the conditions used in deriving the free surface boundary conditions (3.13).

In order to check the validity of (3.25), once again we can systematically nondimensionalise the equations by writing

$$\rho = \rho_s + \varepsilon\rho_0\hat{\rho}, \quad p = p_s + \varepsilon p_0\hat{p}, \quad \mathbf{u} = u_0(\hat{u}, 0, \hat{w}),$$

$x = LX, z = LZ, t = \omega_0^{-1}T$. Here we choose typical values $\rho_0 = \rho_s(0), p_0 = p_s(0)$, and L and u_0 are, as usual, representative length and velocity scales. Now the linearised equations (3.21) and (3.22) are obtained from (3.17) and (3.18) as long as $\omega_0 = u_0/\varepsilon L$. Moreover (3.19) leads to

$$(\rho_s + \varepsilon\rho_0\hat{\rho})\left[\frac{\partial\hat{u}}{\partial T} + \varepsilon\left(\hat{u}\frac{\partial\hat{u}}{\partial x} + \hat{w}\frac{\partial\hat{u}}{\partial z}\right)\right] = -\frac{\varepsilon^2 p_0}{u_0^2}\frac{\partial\hat{p}}{\partial z}$$

and

$$(\rho_s + \varepsilon\rho_0\hat{\rho})\left[\frac{\partial\hat{w}}{\partial T} + \varepsilon\left(\hat{u}\frac{\partial\hat{w}}{\partial x} + \hat{w}\frac{\partial\hat{w}}{\partial z}\right)\right] = -\frac{\varepsilon^2 p_0}{u_0^2}\frac{\partial\hat{p}}{\partial x} - \frac{\varepsilon^2 \rho_0 gL}{u_0^2}\hat{\rho}.$$

Hence, in order to retrieve (3.23) and (3.24), we need

$$p_0 \sim \rho_0 gL \quad \text{and} \quad u_0 \sim \varepsilon\sqrt{gL}.$$

This example again illustrates the importance of our systematic method. We have chosen the scales above in order to justify the use of equation (3.25). However, were we to be modelling sonic boom propagation in the atmosphere, we would be considering wavelengths much shorter than the length scale of the stratification, and this leads to quite a different model, as we will see in the next section.

We can extend the theory to disturbances that vary in three dimensions about the same basic stratified equilibrium solution and the equation for w becomes

$$\frac{\partial^2}{\partial t^2}\left(\frac{\partial^2 w}{\partial x^2} + \frac{\partial^2 w}{\partial y^2} + \frac{\partial^2 w}{\partial z^2}\right) = -N^2(z)\left(\left(\frac{\partial^2 w}{\partial x^2} + \frac{\partial^2 w}{\partial y^2}\right) - g^{-1}\frac{\partial^3 w}{\partial z\,\partial t^2}\right). \quad (3.26)$$

We note that the stratification of the fluid destroys any hope of conservation of vorticity. Even in the linear three-dimensional theory, the only vestige that remains is the following argument. Since, from the generalisations of (3.23) and (3.24)

$$\rho_s(z)\frac{\partial\mathbf{u}}{\partial t} = -\nabla\bar{p} - \bar{\rho}g\mathbf{k},$$

we can deduce that

$$\rho_s\frac{\partial}{\partial t}(\nabla\wedge\mathbf{u}) + \rho_s'\mathbf{k}\wedge\frac{\partial\mathbf{u}}{\partial t} = -g\nabla\bar{\rho}\wedge\mathbf{k}$$

and so

$$\mathbf{k}.\frac{\partial\boldsymbol{\omega}}{\partial t} = 0.$$

Hence the vertical component of the vorticity is conserved in time.

3.4 Acoustic Waves in a Stratified Fluid

As suggested earlier, it is interesting to note what happens when we combine some aspects of the previous section with those of Section 3.1 and consider *acoustic waves* in an inhomogeneous compressible atmosphere so that we have to revert to the full continuity equation $d\rho/dt + \rho\nabla.\mathbf{u} = 0$. For simplicity we neglect the effect of gravity so that $\rho = \rho_s(z)$, but $p_s(z) = $ constant.

The continuity equation linearises to

$$\frac{\partial\bar{\rho}}{\partial t} + \rho_s\nabla.\mathbf{u} + \mathbf{u}.\nabla\rho_s(z) = 0$$

and the momentum equation is just

$$\rho_s(z)\frac{\partial\mathbf{u}}{\partial t} = -\nabla\bar{p}.$$

We now need to close the system with the energy equation $d/dt\,(p/\rho^\gamma) = 0$ which linearises to

$$\frac{1}{p_s}\frac{\partial\bar{p}}{\partial t} = \frac{\gamma}{\rho_s(z)}\left(\frac{\partial\bar{\rho}}{\partial t} + \mathbf{u}.\nabla\rho_s(z)\right).$$

Thus, when we write $\gamma p_s/\rho_s(z) = c_s^2(z)$, we find that the flow is described by a velocity potential ϕ such that $\bar{p} = -\partial\phi/\partial t$ and $\rho_s(z)\mathbf{u} = \nabla\phi$, where

$$\frac{\partial^2\phi}{\partial t^2} = -\frac{\partial\bar{p}}{\partial t} = \gamma p_s\nabla.\left(\frac{1}{\rho_s(z)}\nabla\phi\right) \tag{3.27}$$

$$= \nabla(c_s^2\nabla\phi). \tag{3.28}$$

Note that this result is *not* what we would have obtained by setting $c_0 = c_s(z)$ in (3.6) unless the acoustic wavelength is short compared to the length scale on which c_s varies. If this is the case, we can approximate $\nabla.(\nabla\phi/\rho_s(z))$ by $\nabla^2\phi/\rho_s(z)$ and hence retrieve (3.6). However, for one-dimensional waves where $\phi = \phi(z,t)$, we see that $\int^z \phi(\zeta,t)d\zeta$ does satisfy (3.6) with c_0 replaced by $c_s(z)$. Also we note that although $\bar{p}$ satisfies the same equation as ϕ, the density perturbation $\bar{\rho}$ does not. Note too that c_s^2 is proportional to p_s/ρ_s and thus to the temperature in the undisturbed flow.

If we want to model acoustic waves in an inhomogeneous *ocean*, however, we have to allow p_s to vary with z according to (3.20). As mentioned after (2.13), it is still reasonable to assume that there is a barotropic relation between p and ρ so that $p/\rho^\gamma = p_s/\rho_s^\gamma$, and hence the acoustic variations $\bar{p}$ and $\bar{\rho}$ are linearly related by $\bar{p} = c_s^2\bar{\rho}$, where $c_s^2 = \gamma p_s/\rho_s$. Then, if we also assume that the acoustic wavelength is much smaller than the length scale on which p_s and ρ_s vary, we can again obtain (3.6) to lowest order with c_0^2 replaced by c_s^2.

These examples illustrate how acoustic wave propagation in inhomogeneous fluids can lead to different forms of the wave equation; this also applies outside the realm of fluid mechanics as we will see in Section 3.8.

3.5 Waves in Rotating Incompressible Flows

It can be shown (see Acheson [1]) that the equations of motion of a constant-density inviscid fluid which is moving with velocity $\mathbf{u}$ relative to a set of axes which are rotating with *constant* angular velocity $\boldsymbol{\Omega}$ with respect to a fixed inertial frame are

$$\nabla.\mathbf{u} = 0$$

and

$$\frac{\partial \mathbf{u}}{\partial t} + (\mathbf{u}.\nabla)\mathbf{u} + 2\boldsymbol{\Omega} \wedge \mathbf{u} + \boldsymbol{\Omega} \wedge (\boldsymbol{\Omega} \wedge \mathbf{r}) = -\frac{1}{\rho}\nabla p. \qquad (3.29)$$

Here $\mathbf{r}$ is the position vector, in the rotating frame, of the fluid particle whose velocity in that frame is $\mathbf{u}$ and, most importantly, all spatial derivatives are taken relative to the rotating frame. A simple-minded argument to explain (3.29) is based on the formula that the rate of change of any vector $\mathbf{a}$ with respect to a rotating frame is

$$\frac{d\mathbf{a}}{dt} + \boldsymbol{\Omega} \wedge \mathbf{a}.$$

Hence, the velocity of the particle with position vector $\mathbf{r}$ is

$$\frac{d\mathbf{r}}{dt} + \boldsymbol{\Omega} \wedge \mathbf{r} = \mathbf{u} + \boldsymbol{\Omega} \wedge \mathbf{r},$$

and its acceleration will be

$$\left(\frac{d}{dt} + \boldsymbol{\Omega} \wedge\right)(\mathbf{u} + \boldsymbol{\Omega} \wedge \mathbf{r}) = \frac{d\mathbf{u}}{dt} + 2\boldsymbol{\Omega} \wedge \mathbf{u} + \boldsymbol{\Omega} \wedge (\boldsymbol{\Omega} \wedge \mathbf{r}),$$

and, to account for convection, we must interpret $d/dt = \partial/\partial t + \mathbf{u}.\nabla$. This is a plausible but by no means a watertight argument!

We can simplify (3.29) since $\boldsymbol{\Omega} \wedge (\boldsymbol{\Omega} \wedge \mathbf{r}) = -\nabla(\frac{1}{2}(\boldsymbol{\Omega} \wedge \mathbf{r})^2)$ and so, incorporating a centrifugal term in the pressure leads to

$$\frac{\partial \mathbf{u}}{\partial t} + (\mathbf{u}.\nabla)\mathbf{u} + 2\boldsymbol{\Omega} \wedge \mathbf{u} = -\frac{1}{\rho}\nabla p', \qquad (3.30)$$

where the *reduced pressure* $p' = p - \frac{1}{2}\rho|\boldsymbol{\Omega} \wedge \mathbf{r}|^2$. Now a handwaving linearisation about an equilibrium state $\mathbf{u} = 0$, $p' = p_0$ leads to

$$\frac{\partial \mathbf{u}}{\partial t} + 2\boldsymbol{\Omega} \wedge \mathbf{u} = -\frac{1}{\rho}\nabla p', \qquad (3.31)$$

and a systematic analysis along the lines used in the previous three sections reveals that the nonlinear term in (3.30) can be neglected if the *Rossby Number*, Ro, defined as $U_0/L\Omega$ is small. The systematic analysis also shows that the appropriate time scale for this flow is Ω^{-1}. For meteorological flows on the surface of the earth, we might choose $L = 10^3$ km, $U_0 = 10$ ms^{-1} and, of course, Ω is one revolution per day, so that $Ro \simeq 0.15$. Also, we note that, for a steady flow, (3.31) shows that $\mathbf{u}.\nabla p' = 0$; this explains why the wind velocity is parallel to the isobars on which the reduced pressure is constant as we see daily on weather maps. The term $2\boldsymbol{\Omega} \wedge \mathbf{u}$ in (3.31) is called the *Coriolis term*.

Alas, as in stratified fluids, the flow governed by (3.31) inevitably results in vorticity generation when $\boldsymbol{\Omega} \neq \mathbf{0}$. However, if we take $\boldsymbol{\Omega} = \Omega\mathbf{k}$, it is easy to show from (3.31) that p' and each component of $\mathbf{u}$ all satisfy the inertial wave equation

$$\frac{\partial^2}{\partial t^2}\left(\frac{\partial^2\phi}{\partial x^2} + \frac{\partial^2\phi}{\partial y^2} + \frac{\partial^2\phi}{\partial z^2}\right) = -4\Omega^2\frac{\partial^2\phi}{\partial z^2}. \tag{3.32}$$

As stated by Greenspan [23], "the balance between pressure gradient and Coriolis force emerges as the backbone of the entire subject (of rotating flows)". Already we can see the importance of Ω in determining the frequency of oscillatory solutions of (3.32) and the similarities and differences between this model and the internal wave model given by (3.26).

3.6 *Waves in Dissociating Gases

Gases may sometimes be subject to forces strong enough for the molecules to be split and this phenomenon can be roughly classified as *dissociation* or *ionisation*. In the latter case, the split molecules have positive or negative charge and form a *plasma*, and a simple model for plasmas will be given in Section 3.8.2.

When dissociation occurs, no electromagnetic forces are involved and we can use a rate equation to describe the departure of the molecules from their original chemical state. We can do this most simply by regarding the gas as having two temperatures: its equilibrium temperature T and an *internal temperature* T_i. The temperature T_i measures the energy in the molecular vibrational state of the dissociating gas and is only equal to T in thermodynamic equilibrium. The state furthest from equilibrium is when $T_i = 0$ and this is called the *frozen* state.

As in traditional models for chemical reactions, we postulate a rate equation

$$\tau\frac{dT_i}{dt} = (T - T_i), \tag{3.33}$$

which governs the *relaxation* of the internal temperature to its equilibrium value; τ is the timescale for this relaxation process. The model (3.33) describes a dissipative mechanism (and this is the first time in this book that we have encountered such a situation) which allows energy to be dissipated as the gas changes from the frozen state into the equilibrium state. We also assume that the internal energy is given by

$$e = c_{v_f} T + c_{v_i} T_i, \tag{3.34}$$

where c_{v_f} is the specific heat of the frozen gas and $c_{v_f} + c_{v_i} = c_{v_e}$ is the specific heat in equilibrium. When we use (3.33) and (3.34) with the perfect gas law (2.13) to eliminate T and T_i, we see that

$$\left(\tau\frac{d}{dt} + 1\right) e = \frac{\tau}{\gamma_f - 1}\frac{d}{dt}\left(\frac{p}{\rho}\right) + \frac{1}{\gamma_e - 1}\frac{p}{\rho}, \tag{3.35}$$

where γ_e and γ_f are defined by $c_{v_f} = R/(\gamma_f - 1)$ and $c_{v_e} = R/(\gamma_e - 1)$, so that $\gamma_f > \gamma_e$. Now we can use the energy equation (2.9) with (3.35) to get

$$\left(\tau\frac{d}{dt} + 1\right)\left(\frac{p}{\rho^2}\frac{d\rho}{dt}\right) = \frac{d}{dt}\left(\frac{\tau}{\gamma_f - 1}\frac{d}{dt}\left(\frac{p}{\rho}\right) + \frac{1}{\gamma_e - 1}\frac{p}{\rho}\right),$$

where k and Q are assumed to be zero. After some manipulation, this becomes

$$\rho\bar{\tau}\frac{d}{dt}\left(\frac{1}{\rho}\left(\frac{dp}{dt} - c_f^2\frac{d\rho}{dt}\right)\right) + \frac{dp}{dt} - c_e^2\frac{d\rho}{dt} = 0, \tag{3.36}$$

where $c_f = \sqrt{\gamma_f p/\rho}$ and $c_e = \sqrt{\gamma_e p/\rho}$ are the speeds of sound in the frozen and equilibrium gas and $\bar{\tau} = ((\gamma_e - 1)/(\gamma_f - 1))\,\tau$.

As in conventional gas dynamics, it is easiest to see the general properties of the solution for acoustic waves and we also restrict our discussion here to one dimension for simplicity. Following our usual linearising procedure, we write (2.6) and (2.7) in the form (3.1) and (3.2), and then use (3.36) to get

$$\bar{\tau}\frac{\partial}{\partial t}\left(\frac{\partial^2 u}{\partial t^2} - c_{f0}^2\frac{\partial^2 u}{\partial x^2}\right) + \frac{\partial^2 u}{\partial t^2} - c_{e0}^2\frac{\partial^2 u}{\partial x^2} = 0, \tag{3.37}$$

where c_{f0}, c_{e0} are the undisturbed speeds of sound and $c_{e0} < c_{f0}$.

3.7 *Waves in Viscous Incompressible Flow

The introduction of viscosity into the modelling of a fluid has a dramatic effect. To simplify the discussion, we will assume the fluid to be incompressible[4] throughout this section and then, as shown in [41] or [1], viscosity introduces an extra term

---

[4] To include both viscosity *and* compressibility makes things much more complicated and is described in detail in Stewartson [51].

$\nu \nabla^2 \mathbf{u}$ into the right-hand side of the momentum equation (2.7). Here ν is the kinematic viscosity of the fluid which is assumed to be constant. Thus, in the absence of body forces, the equations of conservation of mass and momentum are

$$\nabla . \mathbf{u} = 0 \tag{3.38}$$

and

$$\frac{\partial \mathbf{u}}{\partial t} + (\mathbf{u}.\nabla)\mathbf{u} = -\frac{1}{\rho}\nabla p + \nu \nabla^2 \mathbf{u}, \tag{3.39}$$

which are the *Navier–Stokes Equations* for an incompressible viscous fluid. The extra term in (3.39) models the fact that the stress within the fluid is due not only to an isotropic pressure but also the frictional *shear stresses*, which are assumed proportional to the velocity gradients.[5] Since ρ is constant, (2.27) is replaced by

$$\frac{d\omega}{dt} = (\omega.\nabla)\mathbf{u} + \nu \nabla^2 \omega$$

and, as we will see in Chapter 4, this shows that viscosity is another damping mechanism that is stronger than the damping associated with dissociation in Section 3.6.

In the simplest case of two-dimensional flow, the continuity equation (3.38) allows us to introduce a *stream function* ψ such that

$$\mathbf{u} = \left(\frac{\partial \psi}{\partial y}, -\frac{\partial \psi}{\partial x}, 0 \right)$$

and $\omega = (0, 0, \omega)$. Then, our equation for the vorticity can be written as

$$\frac{\partial \omega}{\partial t} + \frac{\partial(\psi, \omega)}{\partial(y, x)} = \nu \nabla^2 \omega, \tag{3.40}$$

where

$$\omega = -\nabla^2 \psi. \tag{3.41}$$

Even these equations are a formidable nonlinear system, but they admit solutions in which the flow is steady and unidirectional when $\psi = \psi_0(y)$ is a cubic function of y. Then, when we write $\psi = \psi_0(y) + \tilde{\psi}$ and neglect squares of $\tilde{\psi}$, we find the following linear evolution equation for $\tilde{\psi}$:

$$\left[\frac{\partial}{\partial t} + \psi_0'(y)\frac{\partial}{\partial x} - \nu \left(\frac{\partial^2}{\partial x^2} + \frac{\partial^2}{\partial y^2} \right) \right] \left(\frac{\partial^2 \tilde{\psi}}{\partial x^2} + \frac{\partial^2 \tilde{\psi}}{\partial y^2} \right) = \psi_0'''(y)\frac{\partial \tilde{\psi}}{\partial x}, \tag{3.42}$$

which is known as an *Orr–Sommerfeld equation*.

---

[5] To be more precise, the stress in the fluid is a tensor, τ_{ij}, which is linearly related to the rate-of-strain tensor $\partial u_i / \partial x_j$, as described in Chapter 1 of [41].

3.8 *Other Linear Wave Models in Mechanics and Electromagnetism

There are many continuum mathematical models that lead to linear wave equations different from those encountered in fluid mechanics. In this section, we have chosen four examples that illustrate the diversity of these equations and the breadth of the applicability of linear wave theory. The fundamental models for the last three phenomena described are nonlinear, as was the case with water waves and gas dynamics, but here we will only give brief details of their derivation before carrying out the linearisation.

3.8.1 Electromagnetic Waves

Electromagnetic theory is based on *Maxwell's equations* which are linear and deceptively simple. They simply state that, in free space, the electric field $\mathbf{E}$ and the magnetic field $\mathbf{H}$ are related by

$$\nabla \wedge \mathbf{H} = \varepsilon \frac{\partial \mathbf{E}}{\partial t} \tag{3.43}$$

and

$$\nabla \wedge \mathbf{E} = -\mu \frac{\partial \mathbf{H}}{\partial t}, \tag{3.44}$$

where ε, μ are positive constants[6] and $\nabla.\mathbf{E} = \nabla.\mathbf{H} = 0$. Unfortunately, an explanation of these equations can take many pages, but a simple derivation is described in Coulson and Boyd [12]. For our purposes, the principal result is that, by cross-differentiation, all the components of $\mathbf{E}$ and $\mathbf{H}$ satisfy the wave equation

$$\nabla^2 \phi = \frac{1}{c^2} \frac{\partial^2 \phi}{\partial t^2}, \tag{3.45}$$

where $c = 1/\sqrt{\varepsilon\mu}$ is now the speed of light.

Maxwell's equations are a prototype for modelling new phenomena such as polarisation as we shall see in Section 3.8.3. Moreover, they can be written in the conservation form (2.10) in two interesting ways. First we can combine (3.43) and (3.44) as

$$\frac{\partial}{\partial t} \left\{ \frac{1}{2}\varepsilon|\mathbf{E}|^2 + \frac{1}{2}\mu|\mathbf{H}|^2 \right\} + \nabla.(\mathbf{E} \wedge \mathbf{H}) = 0,$$

where we can interpret the term in curly brackets as an energy density and $\mathbf{E} \wedge \mathbf{H}$ (which is called the *Poynting vector*) as an energy flux.

---

[6] They are matrices in an anisotropic medium.

Another conservation equation arises from the same equations (3.43) and (3.44) when we note that

$$\mu\varepsilon\frac{\partial}{\partial t}(\mathbf{E} \wedge \mathbf{H}) + \{\varepsilon\mathbf{E} \wedge (\nabla \wedge \mathbf{E}) + \mu\mathbf{H} \wedge (\nabla \wedge \mathbf{H})\} = \mathbf{0},$$

where the term in curly brackets is the divergence of the *Maxwell stress tensor* T_{ij}, where

$$T_{ij} = \varepsilon E_i E_j + \mu H_i H_j - \frac{1}{2}\delta_{ij}(\varepsilon E_k E_k + \mu H_k H_k).$$

Maxwell's equations can be extended to include charge and current as follows:

(i) When charges are present the electric field no longer has zero divergence and instead

$$\varepsilon\nabla.\mathbf{E} = \rho, \tag{3.46}$$

where ρ is the charge density.

(ii) If we are considering electromagnetic fields in an electrical conductor such as a metal, then (3.43) is modified to become

$$\nabla \wedge \mathbf{H} = \varepsilon\frac{\partial \mathbf{E}}{\partial t} + \mathbf{j}, \tag{3.47}$$

where $\mathbf{j}$ is the current density in the conductor. Note that for a steady current in a wire, this equation reduces to the *Biot–Savart law*. Since we have introduced a new dependent variable $\mathbf{j}$, the system now needs to be closed by a constitutive law which is usually taken to be Ohm's law

$$\mathbf{j} = \sigma\mathbf{E}, \tag{3.48}$$

where σ is called the conductivity of the material. As we shall see in Section 4.4, this formulation shows that conductors will inevitably create dissipation of an electromagnetic field.

3.8.2 Waves in Plasmas

When matter becomes sufficiently hot, or is subject to a sufficiently strong electromagnetic field (as in the sun or in a discharge lamp), the ions and electrons can move around fairly freely (together with neutral particles) to form a plasma. Under these circumstances, the "fluid particles" cannot realistically be modelled deterministically, as in Chapter 2, and ideas from statistical mechanics must be employed.

Instead of thinking of a particle as having a well-defined position $\mathbf{x}$ and velocity $\mathbf{u}$ at any particular time, we have to work with probability distribution functions which depend on $\mathbf{x}$, t and the *random variable* $\mathbf{u}$; $\mathbf{x}$, $\mathbf{u}$ are then vectors in *phase space*.

Then, if the motion is on average in just the x-direction (which is the only case we will consider), we define $f(x, u, t)\delta x \delta u$ to be the probability that a particle at time t lies in the small region $\delta x \delta u$ of phase space (x, u).

The only ways in which the different species, that is, the electrons, ions and neutrals, can interact is

(i) via the electromagnetic field that they generate or that is externally imposed, and

(ii) via collisions which occur when the separation between the particles is small enough to cause a local change in their trajectories.

In the fluid configurations considered in Chapter 2, the charged particles (ions and electrons) remain attached to the fluid molecules and it is the collisions between these molecules that ultimately generate the pressure in the continuum model of the fluid. However, at the opposite extreme, when there are many free charged particles, they rarely collide. Indeed, in a fluorescent light bulb the "mean free path", which is the mean distance between collisions, may be of the order of centimetres. Thus, in a plasma we assume that, in the absence of particle generation or annihilation, the distribution function f is conserved as we follow a particle in phase space. Hence, generalising (2.6), we can write

$$\frac{\partial f}{\partial t} + u\frac{\partial f}{\partial x} + \frac{F}{m}\frac{\partial f}{\partial u} = 0, \tag{3.49}$$

where, in the same way that u is the rate of change of x following a particle, F/m is the rate of change of u, F being the force on a particle and m its mass. Equation (3.49) is called the *Vlasov equation* for the species under consideration. We comment that, although the random velocities are inevitably three dimensional, if the net velocity is in the x-direction, then the three-dimensional version of the Vlasov equation can be averaged in the y and z directions to give (3.49). For a more detailed discussion of plasma modelling, see [10].

We will only consider plasmas which consist of ions with positive charge e and electrons with negative charge $-e$. We suppose that the ions and electrons have distribution functions $f_i(x, u_i, t)$ and $f_e(x, u_e, t)$, respectively. Fortunately, in many practical situations it is the electric field $\mathbf{E}$ which provides the force on the particles and the effects of the magnetic field $\mathbf{H}$ can be neglected. In this case, (3.44) implies that $\mathbf{E} = -\nabla\phi$, where ϕ is the electric potential. Then, from (3.46) we get

$$\varepsilon\nabla^2\phi = e(n_e - n_i), \tag{3.50}$$

where n_e and n_i are the number densities of electrons and ions, respectively, which, by definition of the distribution function, satisfy

$$n_e = \int_{-\infty}^{\infty} f_e(x, u, t)\, du, \quad n_i = \int_{-\infty}^{\infty} f_i(x, u, t)\, du. \tag{3.51}$$

3.8.2.1 Ion Waves

We first consider the case in which the amplitude of the rapidly oscillating thermal velocity of the electrons is much greater than their average *drift velocity*, defined to be $\bar{u}_e(x, t) = \int_{-\infty}^{\infty} u_e f_e(x, u_e, t) du_e / n_e$. In this case, the electron collisions cause them to effectively behave like a gas which can be modelled as in Chapter 2. Thus, we no longer use the Vlasov equation for f_e but instead assume that this electron–gas has a pressure p_e and a temperature T_e which satisfy the gas law (2.13) in the form

$$p_e = n_e k T_e,$$

where k is proportional to R (although k and R have different dimensions). Then, in local equilibrium under an electric field $\mathbf{E} = -(\partial \phi / \partial x)\, \mathbf{i}$, the force balance on the electrons is

$$\frac{\partial p_e}{\partial x} = e n_e \frac{\partial \phi}{\partial x},$$

so that, if we assume T_e is constant, we can integrate to get

$$n_e = n_0 \exp\left(\frac{e\phi}{kT_e}\right), \tag{3.52}$$

where n_0 is a constant. Now we can nondimensionalise (3.49)–(3.52) and rearrange to get the ion-wave equations

$$\frac{\partial f_i}{\partial t} + u_i \frac{\partial f_i}{\partial x} - \frac{\partial \phi}{\partial x} \frac{\partial f_i}{\partial u_i} = 0, \tag{3.53}$$

$$\frac{\partial^2 \phi}{\partial x^2} = e^\phi - n_i \tag{3.54}$$

and

$$n_i = \int_{-\infty}^{\infty} f_i(x, u, t)\, du, \tag{3.55}$$

which must be solved subject to suitable initial and boundary conditions.

The simplest case to consider is the so-called *cold-ion limit* in which the ion velocity fluctuations are negligible compared to their drift velocity $\bar{u}(x, y)$. Thus, we can write

$$f_i = n_i(x, t)\delta(u_i - \bar{u}(x, t)),$$

where δ is the Dirac delta function.[7] Then, by integrating (3.53) with respect to u_i (see Exercise 3.13), we obtain

$$\frac{\partial n_i}{\partial t} + \frac{\partial}{\partial x}(n_i \bar{u}) = 0. \tag{3.56}$$

---

[7] For a definition of the Dirac delta function see, for instance, [43].

Similarly, if we first multiply by u_i and then integrate (3.53), after using (3.56) we have

$$\frac{\partial \bar{u}}{\partial t} + \bar{u}\frac{\partial \bar{u}}{\partial x} = -\frac{\partial \phi}{\partial x}. \tag{3.57}$$

Thus, the ions behave like a barotropic gas with $\phi = \int dp/\rho$ but now the "equation of state" is given by (3.54). We note the following special cases:

(a) When the term $\partial^2 \phi/\partial x^2$ is negligible, (3.54) gives a gas law $n_i = e^\phi$ which corresponds to $\gamma = 1$.
(b) By linearising about the exact solution of (3.54), (3.56) and (3.57) in which $n_i = 1$, $\phi = \bar{u} = 0$, we obtain

$$\frac{\partial^2 \phi}{\partial x^2} = \phi - \tilde{n}, \tag{3.58}$$

$$\frac{\partial \tilde{n}}{\partial t} + \frac{\partial \bar{u}}{\partial x} = 0 \tag{3.59}$$

and

$$\frac{\partial \bar{u}}{\partial t} + \frac{\partial \phi}{\partial x} = 0, \tag{3.60}$$

where $n_i = 1 + \tilde{n}$. These three equations describe *ion-acoustic waves* and by eliminating $\tilde{n}$ and $\bar{u}$ we can write them as a single fourth-order equation

$$\frac{\partial^4 \phi}{\partial x^2 \partial t^2} = \frac{\partial^2 \phi}{\partial t^2} - \frac{\partial^2 \phi}{\partial x^2}. \tag{3.61}$$

3.8.2.2 Electron Waves

Another relatively simple situation occurs when the ions are virtually immobile so that $u_i = 0$ and n_i is constant. Now, if the electrons are not too hot, we do need to consider the Vlasov equation (3.49) for the electrons and the nondimensional forms of (3.49)–(3.51) for f_e, n_e, ϕ are

$$\frac{\partial f_e}{\partial t} + u_e\frac{\partial f_e}{\partial x} + \frac{\partial \phi}{\partial x}\frac{\partial f_e}{\partial u_e} = 0 \tag{3.62}$$

and

$$\frac{\partial^2 \phi}{\partial x^2} = n_e - n_i, \tag{3.63}$$

where n_i is the dimensionless ion density and, again

$$n_e = \int_{-\infty}^{\infty} f_e(x, u, t)\, du. \tag{3.64}$$

We note that if we model the "cold-electron" limit as we did for cold ions by writing $f_e = n_e\delta(u_e - \bar{u})$ and putting $n_i = 1$, equations (3.62)–(3.64) reduce to

$$\frac{\partial n_e}{\partial t} + \frac{\partial}{\partial x}(n_e \bar{u}) = 0,$$

$$\frac{\partial \bar{u}}{\partial t} + \bar{u}\frac{\partial \bar{u}}{\partial x} - \frac{\partial \phi}{\partial x} = 0$$

$$\frac{\partial^2 \phi}{\partial x^2} = n_e - 1,$$

after integrating (3.62) with respect to u_e as for the cold-ion limit above. If we now linearise about $\bar{u} = 0$ and $n_e = 1$, writing $n_e = 1 + \hat{n}$, we find that *electron-acoustic waves* satisfy

$$\frac{\partial \hat{n}}{\partial t} + \frac{\partial \bar{u}}{\partial x} = 0, \tag{3.65}$$

$$\frac{\partial \bar{u}}{\partial t} - \frac{\partial \phi}{\partial x} = 0 \tag{3.66}$$

and

$$\frac{\partial^2 \phi}{\partial x^2} - \hat{n} = 0. \tag{3.67}$$

More generally we can study situations in which the electrons have non-zero temperature by linearising equations (3.62)–(3.64) about the steady solution in which $f_e = f_0(u_e)$, $\phi = 0$, $n_e = n_i$, so that

$$\frac{\partial \tilde{f}}{\partial t} + u_e \frac{\partial \tilde{f}}{\partial x} + \frac{\partial \phi}{\partial x} f_0'(u_e) = 0, \tag{3.68}$$

$$\frac{\partial^2 \phi}{\partial x^2} = \tilde{n} \tag{3.69}$$

and

$$\tilde{n} = \int_{-\infty}^{\infty} \tilde{f}(x, u, t)\, du, \tag{3.70}$$

where $n_e = n_i + \tilde{n}$ and $f_e = f_0(u_e) + \tilde{f}(x, u_e, t)$.

A common assumption is to take $f_0(u_e)$ to be a thermodynamic equilibrium distribution, which is a Gaussian. This follows from the fact that f_0 satisfies Boltzmann's equation, which is the steady form of the Vlasov equation (3.49), but with an integral term on the right-hand side which models collisions. It can be shown that a stable equilibrium solution of Boltzmann's equation is the Maxwellian distribution

$$f_0(u_e) = \frac{n_e \sqrt{m}}{\sqrt{2\pi k T_e}} \exp\left(-\frac{m u_e^2}{2k T_e}\right), \tag{3.71}$$

where m is the mass of an electron and T_e is its temperature [8]. We shall see in Chapter 4 that the systems (3.65)–(3.67) and (3.68)–(3.70) have very different solutions from those described by (3.61).

3.8.3 Elastic Waves

The most familiar models for elastic waves are those for small transverse vibrations of a string or membrane of density ρ under tension T; the displacement simply satisfies the one- or two-dimensional version of the wave equation (3.6) with $c_0^2 = T/\rho$. However, the equation for small-amplitude waves in linear elastic solids looks a little more formidable and can be written as

$$\rho\frac{\partial^2 \mathbf{X}}{\partial t^2} = (\lambda + 2\mu)\nabla(\nabla.\mathbf{X}) - \mu\nabla \wedge (\nabla \wedge \mathbf{X}), \qquad (3.72)$$

where $\mathbf{X}$ is the displacement of an element of the material from its equilibrium position $\mathbf{a}$, where we are now working in *Lagrangian coordinates*, so that $\nabla = (\partial/\partial a_1, \partial/\partial a_2, \partial/\partial a_3)$. The density ρ is assumed constant and parameters λ and μ are called the Lamé constants[8] of the material. As may be guessed, this equation represents conservation of momentum and has some similarities with the Navier–Stokes equation for viscous flow.

The derivation of (3.72) is difficult to explain without the use of two tensors.

(i) The stress tensor, τ_{ij}, is defined as the force per unit area in the a_i direction acting on a surface element whose normal is the a_j direction.

(ii) The strain tensor, $e_{ij} = \dfrac{1}{2}\left(\dfrac{\partial X_i}{\partial a_j} + \dfrac{\partial X_j}{\partial a_i}\right)$.

Then, as described in [28], these tensors are related in linear elasticity theory by the generalised Hooke's law

$$\tau_{ij} = \lambda\delta_{ij}e_{kk} + 2\mu e_{ij}, \qquad (3.73)$$

where e_{kk} is the trace of e_{ij} and δ_{ij} is the Kronecker delta. Now, (3.72) follows directly from Newton's law

$$\rho\frac{\partial^2 X_i}{\partial t^2} = \frac{\partial \tau_{ij}}{\partial a_j}. \qquad (3.74)$$

It can be seen that the term involving λ represents the stresses that do work in expansion or compression (like the pressure term in (2.7)) whereas the terms in μ represent the shear stresses and are analogous to the viscous terms in the Navier–Stokes equation (3.39).

The mathematical structure of (3.72) can be discerned by noticing that if $\mathbf{X} = \nabla\phi$, then

$$\nabla^2\phi = \frac{1}{c_p^2}\frac{\partial^2\phi}{\partial t^2}, \qquad (3.75)$$

where $c_p^2 = (\lambda + 2\mu)/\rho$, and if $\mathbf{X} = \nabla \wedge \mathbf{w}$, then the components of $\mathbf{w}$ all satisfy

$$\nabla^2 w_i = \frac{1}{c_s^2}\frac{\partial^2 w_i}{\partial t^2}, \qquad (3.76)$$

---

[8] These numbers need to be replaced by matrices for anisotropic materials.

where $c_s^2 = \mu/\rho$ and $c_s < c_p$. As was the case in Section 3.6, there are two distinct wave speeds, c_p and c_s, and we anticipate that general solutions of (3.72) will contain both types of waves.

We also note that, since (3.72) is just an expression of Newton's laws for an elastic material, it does not need to be complemented by an energy equation unless the effects of thermal expansion need to be taken into account. This would lead us to the difficult subject of thermoelasticity which we will not discuss here. However, energy is an important concept for *hyperelastic* materials which are characterised by the constitutive law $\tau_{ij} = \partial W/\partial e_{ij}$ for some suitable *strain energy function W* (see [28]). For our "Hookean" material

$$ W = \frac{1}{2}\lambda(e_{kk})^2 + \mu e_{ij}e_{ij}, $$

and this leads to (3.73) for τ_{ij}. In general, by integrating (3.72) it can be shown that the rate of change of the *total* energy in a volume V,

$$ \frac{d}{dt}\left\{ \iiint_V \left(\frac{1}{2}\rho \left(\frac{\partial \mathbf{X}}{\partial t} \right)^2 + W \right) d\mathbf{X} \right\}, $$

is the work done by any external forces acting on the boundary of V.

Since the field equation for $\mathbf{X}$, like (3.43) and (3.44) for $\mathbf{E}$ and $\mathbf{H}$ in Section 3.8.1, is a vector wave equation, this is a good place to introduce the ideas of *transverse* and *longitudinal* wave motion. In general, waves satisfying vectorial wave equations are *longitudinal* if the vector variable is parallel to the direction of wave motion and *transverse* if it is perpendicular to the direction of the wave. In elasticity, a simple travelling wave solution of (3.75) is $\phi = f(\mathbf{x}.\mathbf{k}_1 - c_p t)$, where $\mathbf{k}_1$ is a fixed unit vector in the wave direction and $\mathbf{X} = \nabla\phi = \mathbf{k}_1 f'(\mathbf{x}.\mathbf{k}_1 - c_p t)$, which is therefore an example of a longitudinal wave. The same solution shows that acoustic waves which satisfy (3.6) are always longitudinal since the velocity is in the direction in which the wave travels. On the other hand, if, in (3.76), we take $\mathbf{w} = \mathbf{k}_2 f(\mathbf{x}.\mathbf{k}_1 - c_s t)$ where $\mathbf{k}_2$ is another constant unit vector, then $\mathbf{X} = \nabla f \times \mathbf{k}_2 = f'(\mathbf{x}.\mathbf{k}_1 - c_s t)\mathbf{k}_1 \times \mathbf{k}_2$ and hence this is a transverse wave. Similarly, in electromagnetics there are transverse wave solutions of (3.43), (3.44) of the form

$$ \mathbf{E} = \mathbf{k}_2 f(\mathbf{x}.\mathbf{k}_1 - ct), \mathbf{H} = \frac{\mathbf{k}_3}{\mu c}f(\mathbf{x}.\mathbf{k}_1 - ct), $$

where $\mathbf{k}_1, \mathbf{k}_2, \mathbf{k}_3$ form an orthonormal triad of vectors. In practice, the functions f in all these examples are usually taken to be complex exponentials so that general solutions can be found by Fourier superposition.

The classification of waves into longitudinal and transverse leads to the more general concept of *polarisation*[9] in vectorial wave equations. For any such equation with constant coefficients, we may seek solutions of the form

---

[9] This is not to be confused with *magnetic polarisation* which is an important phenomenon in electromagnetic theory (see Coulson and Boyd [12]).

$$\mathbf{X} = f(\mathbf{k}.\mathbf{x} - ct)\mathbf{U},$$

where $\mathbf{k}$ is a unit vector in the direction of the wave and $\mathbf{U}$ is a unit vector which depends on the choice of $\mathbf{k}$. A particular solution of this type is called a *plane polarised* wave with $\mathbf{U}$ being the *direction* of polarisation and $(\mathbf{U}, \mathbf{k})$ defining the *plane* of polarisation (assuming it is not a longitudinal wave). As we will see in the next chapter, the general solution can always be written in principle in terms of Fourier integrals as

$$\mathbf{X} = \iiint e^{i(\mathbf{k}.\mathbf{x} - ct)} \mathbf{A}(\mathbf{k})\, d\mathbf{k};$$

polarised waves correspond to the vector $\mathbf{A}$ being "localised" near a particular vector $\mathbf{k}$.

3.8.4 Plastic Waves

We have already seen that waves in fluids can be dissipated by dissociative or viscous mechanisms. It is similarly true that when solids are viscoelastic, so that they have both fluid and elastic components to their stresses, then dissipation can also occur [28]. But granular solids and metals[10] can be subject to a much more dramatic dissipative phenomenon known as *plasticity*. Plastic deformations happen when some measure of the *deviatoric stress* (i.e. the actual stress τ_{ij} minus the pressure $p\delta_{ij}$ in the solid) exceeds a critical value called the *yield stress*. What is really dramatic about plasticity is that it allows the material to undergo a bulk flow, which is the only kind of deformation that a fluid can sustain. Moreover experimental evidence suggests that, at high enough stresses (perhaps 100 times the yield stress), metals can flow exactly like a barotropic gas!

We will consider the effect of plasticity on the compression of a metal slab which is much thinner than it is wide so that the displacement is only in the direction normal to the slab; such a deformation is called *uniaxial strain*. We will begin by considering how to combine the effects of plasticity and elasticity for small displacements and then (although it does not really belong in this chapter) we will derive a model for violent compression which is inevitably nonlinear.

3.8.4.1 Linear Elastoplastic Waves

As discussed in [28], for small strains it is plausible to linearly decompose the strain e_{ij} into elastic and plastic parts,

$$e_{ij} = e^e_{ij} + e^p_{ij},$$

---

[10] In granular materials and metals, plastic flow results from interparticle slip and dislocation motion, respectively; dislocations are line defects in the atomic lattice that cause singularities in stress and strain.

where $\partial e_{ij}^p / \partial t$ is the *rate of strain* tensor which describes the plastic flow. Assuming the critical yield stress has been attained so that plastic flow occurs, the simplest *uniaxial* displacement is modelled by writing the Lagrangian displacement $\mathbf{X} = (X(a,t), 0, 0)$. The stress and strain tensors will be diagonal so that we write

$$\tau_{ij} = \mathrm{diag}(\tau_1(a,t), \tau_2(a,t), \tau_2(a,t)),$$
$$e_{ij}^e = \mathrm{diag}(e_1^e(a,t), e_2^e(a,t), e_2^e(a,t))$$

and

$$e_{ij}^p = \mathrm{diag}(e_1^p(a,t), e_2^p(a,t), e_2^p(a,t)).$$

In addition, the stress will satisfy a yield condition wherever there is plastic flow. Usually, the yield condition is taken to be a measure of the maximum shear stress experienced by the material which in this case can be shown to be $|\tau_1 - \tau_2|$. Thus, we assume that

$$|\tau_1 - \tau_2| = \sigma_Y \tag{3.77}$$

whenever the material is flowing plastically.

To close the model, we also need to assume a *flow rule* which relates τ_{ij} to $\partial e_{ij}^p / \partial t$. This is a contentious modelling issue, but it can be shown that the rate of energy dissipation, which is a measure of the damping rate, induced by plasticity is maximised if $\partial e_{ij}^p / \partial t$ is proportional to the deviatoric stress $\{\tau_{ij} - \frac{1}{3}\tau_{kk}\delta_{ij}\}$; the constant of proportionality is the Lagrangian multiplier in the maximisation. In addition, dislocation theory predicts that the contribution to the stress from e_{ij}^p is negligible and so we assume that τ_{ij} only depends on e_{ij}^e.

In summary, our uniaxial model for elastoplasticity under small strains consists of the following linear equations. The constitutive relations come from linear elasticity (3.73) and are thus

$$\tau_{11} = \tau_1 = (\lambda + 2\mu)e_1^e + 2\lambda e_2^e \tag{3.78}$$

and

$$\tau_{22} = \tau_{33} = \tau_2 = \lambda e_1^e + 2(\lambda + \mu)e_2^e. \tag{3.79}$$

Then, the momentum equation (3.74) is now written as

$$\rho_0 \frac{\partial^2 X}{\partial t^2} = \frac{\partial \tau_1}{\partial a} \tag{3.80}$$

to emphasise that ρ_0 is a constant and the yield criterion is, assuming for definiteness that $\tau_1 > \tau_2$,

$$\tau_1 - \tau_2 = \sigma_Y. \tag{3.81}$$

Finally, the flow rule is

$$\frac{\partial e_1^p}{\partial t} = \frac{2\Lambda}{3}(\tau_1 - \tau_2) \tag{3.82}$$

and

$$\frac{\partial e_2^p}{\partial t} = \frac{\Lambda}{3}(\tau_2 - \tau_1), \tag{3.83}$$

where the Lagrangian multiplier, Λ, is a positive scalar function of a, t and the factor 2 results from uniaxiality. From (3.82) and (3.83), we can immediately see that

$$e_1^p + 2e_2^p = 0,$$

and this implies that the plastic flow is incompressible. From the definition of e_{ij}, we also see that in uniaxial flow

$$\frac{\partial X}{\partial a} = e_1^e + e_1^p \tag{3.84}$$

and

$$0 = e_2^e + e_2^p. \tag{3.85}$$

If an increasing stress or strain is applied to the material, there will initially be a purely elastic response governed by (3.80) with $\tau_1 - \tau_2 < \sigma_Y$ and $e_1^p = e_2^p = 0$. Once the yield criterion is attained, the material will deform plastically, and we have to solve equations (3.78)–(3.85) as well as determining the position of the elastic-plastic boundary in (a, t) space. It can be shown (see Exercise 3.15) that these equations lead to the wave equation

$$\frac{\partial^2 X}{\partial t^2} = \frac{3\lambda + 2\mu}{3\rho_0} \frac{\partial^2 X}{\partial a^2}; \tag{3.86}$$

the positivity of $3\lambda + 2\mu$ can be demonstrated on thermodynamic grounds, even though λ, but not μ, may be negative. Note that the wave speed for these longitudinal plastic waves is $\sqrt{(3\lambda + 2\mu)/3\rho_0}$ which is always slower than the wave speed $c_p = \sqrt{(\lambda + 2\mu)/\rho_0}$ that was found for longitudinal elastic waves in Section 3.8.3.

We conclude by noting that even in this dissipative system, we have not needed to write down an energy equation to close the model. However, it is quite easy to do so, and hence calculate the temperature rise due to the plastic flow. As shown in [28], the dissipation function is $\Phi = \dfrac{\partial e_{ij}}{\partial t}\tau_{ij}$ which is the rate at which mechanical energy is lost and heat is produced. We find that only e_1^p and e_2^p contribute to the result and

$$\Phi = \frac{2\Lambda}{3}(\tau_1^2 + \tau_2^2)$$

in our uniaxial case. In many practical situations as much as 70% of the work put into an elastoplastic deformation is turned into thermal energy.

3.8.4.2 Nonlinear Elastoplastic Waves

Although we have used Lagrangian coordinates in Sections 2.1, 3.8.3 and 3.8.4.1, there is in fact no difference between the Eulerian and Lagrangian formulations when we are considering a linearised problem. However, for larger displacements it is essential that we work with the Lagrangian variables that label the individual material particles with their initial position $\mathbf{a}$, and we cannot avoid the "geometric" nonlinearity inherent in the fact that the deformation $|\mathbf{x} - \mathbf{a}|$ is comparable in size to $|\mathbf{a}|$. We will consider the same uniaxial problem as we did for linear waves.

It was shown in Exercise 2.1 that $d(\rho J)/dt = 0$, where $J = \partial\mathbf{x}/\partial\mathbf{a}$. Hence, in the uniaxial case,

$$\rho\frac{\partial x}{\partial a} = \rho_0, \tag{3.87}$$

which is a statement of mass conservation. Instead of linear strains e_{ij}, we now work with the nonlinear *deformation gradient* tensor

$$F_{ij} = \frac{\partial x_i}{\partial a_j},$$

and we use a multiplicative decomposition[11] to write

$$F_{ij} = F_{ij}^e F_{ij}^p. \tag{3.88}$$

In the uniaxial situation, we can write

$$F_{ij} = \text{diag}\left(\frac{\partial x}{\partial a}, 1, 1\right),$$

$$F_{ij}^e = \text{diag}(F_1^e, F_2^e, F_2^e)$$

and

$$F_{ij}^p = \text{diag}(F_1^p, F_2^p, F_2^p),$$

so that

$$\frac{\partial x}{\partial a} = F_1^e F_1^p = \frac{\rho_0}{\rho} \tag{3.89}$$

and

$$F_2^e F_2^p = 1. \tag{3.90}$$

The components of F_{ij} are the *stretches* that measure the lengths of line segments in the deformed material relative to their original lengths. Note that we can retrieve the linear theory by writing $F_i = 1 + e_i$, where $|e_i| \ll 1$.

---

[11] This is yet another contentious issue in plasticity modelling; we have chosen "p" before "e" purely on physical grounds, but the ordering is irrelevant in a uniaxial configuration.

The momentum equation (3.80), with $X = x - a$, and the yield criterion (3.81) will still hold. However, the nonlinear generalisations of (3.82) and (3.83) become

$$\frac{1}{F_1^p}\frac{\partial F_1^p}{\partial t} = \frac{2\Lambda}{3}(\tau_1 - \tau_2)$$

and

$$\frac{1}{F_2^p}\frac{\partial F_2^p}{\partial t} = \frac{\Lambda}{3}(\tau_2 - \tau_1).$$

Thus, we can deduce that

$$F_1^p(F_2^p)^2 = 1, \tag{3.91}$$

which again implies that the material is incompressible in the plastic phase.

The linear stress-strain relations (3.78) and (3.79) are no longer appropriate and so we need a new constitutive relation between τ_i and F_i^e to complete the model. As in Section 3.8.3, we assume that τ_i can be derived from a strain energy function $W(F_1^e, F_2^e)$, and for simplicity, we will continue to assume linear "mechanical" elasticity. As shown in Exercise 3.16, this assumption enables us to deduce formulae for τ_1 and τ_2 in terms of F_1^e, F_2^e and ρ. Then, by manipulating the algebraic relations (3.81), (3.89), (3.90) and (3.91), we ultimately end up with τ_1 as a function of ρ. Finally, we restrict ourselves to the case of *violent compression* in which the applied stress is much larger than the yield stress σ_Y. Then, (3.81) implies that $\tau_1 \simeq \tau_2$ in the plastic flow and so we see that the stress tensor is approximately $-p\delta_{ij}$, where $\tau_1 = \tau_2 = -p$, and the material behaves exactly like an inviscid compressible barotropic fluid where the pressure depends on the density [30]. When we solve explicitly for p as a function of ρ, we find that

$$p = -\tau_1 = (3\lambda + 2\mu)\left(\frac{\rho}{\rho_0}\right)^{1/3}\left(\left(\frac{\rho}{\rho_0}\right)^{1/3} - 1\right). \tag{3.92}$$

As we will see in Chapters 4 and 6, this model leads to waves that are quite unlike the small-amplitude plastic waves generated by the model (3.86) derived earlier in this section. Moreover the fact that (3.92) is a non-convex function of ρ means that we can, when we revert to Eulerian coordinates, expect behaviour quite different from the conventional barotropic gas dynamics that we will encounter in Chapter 5.

Exercises

R 3.1 (i) Show that, in three dimensions, the linearised equations for acoustic flow, namely, (3.1) and (3.2), are replaced by

$$\frac{\partial \bar{\rho}}{\partial t} + \rho_0 \nabla . \mathbf{u} = 0$$

and

$$\frac{\partial \mathbf{u}}{\partial t} + \frac{1}{\rho_0} \nabla \bar{p} = \mathbf{0}$$

and deduce that $\bar{p}, \bar{\rho}$ and $\mathbf{u}$ all satisfy equation (3.6).

(ii) Suppose now that $\bar{\mathbf{u}} = \mathbf{u} - U\mathbf{i}$ is small, in addition to $\bar{p}$ and $\bar{\rho}$. Show that the linearised equations are then

$$\left(\frac{\partial}{\partial t} + U \frac{\partial}{\partial x} \right) \bar{\rho} + \rho_0 \nabla . \bar{\mathbf{u}} = 0$$

and

$$\left(\frac{\partial}{\partial t} + U \frac{\partial}{\partial x} \right) \bar{\mathbf{u}} + \frac{1}{\rho_0} \nabla \bar{p} = \mathbf{0}$$

and deduce that $\bar{p}, \bar{\rho}$ and $\bar{\mathbf{u}}$ all satisfy

$$\nabla^2 \phi = \frac{1}{c_0^2} \left(\frac{\partial}{\partial t} + U \frac{\partial}{\partial x} \right)^2 \phi.$$

Show that this reduces to (3.6) if we change to moving axes (X, y, z), where $X = x - Ut$ and we assume that U is constant.

R 3.2 Gas is contained in a box $0 < x < A$, $0 < y < B$, $0 < z < C$. Show that acoustic oscillations satisfying (3.6) are possible in which ϕ is proportional to $\cos \omega t$ if

$$\omega^2 = \pi^2 c_0^2 \left(\frac{l^2}{A^2} + \frac{m^2}{B^2} + \frac{n^2}{C^2} \right),$$

where l, m, n are integers.

Show also that if just one face of the box is subject to small-amplitude oscillations so that

$$\frac{\partial \phi}{\partial x} = a \cos \omega t$$

on $x = 0$, then, in general, a possible solution is

$$\phi = \frac{ac_0 \cos\left[\omega(A - x)/c_0\right]}{\omega \sin \omega A/c_0} \cos \omega t.$$

For what values of ω is this solution inadmissible?
Show that if $A = \infty$ and

$$\frac{\partial \phi}{\partial x} = a \cos \frac{m\pi y}{B} \cos \frac{n\pi z}{C} \cos \omega t$$

on $x = 0$, then, if $m^2/B^2 + n^2/C^2 > \omega^2/c_0^2\pi^2$, there are solutions of the form

$$\phi = -\frac{a}{\lambda} \cos \frac{m\pi y}{B} \cos \frac{n\pi z}{C} e^{-\lambda x} \cos \omega t,$$

where $\lambda^2 = m^2\pi^2/B^2 + n^2\pi^2/C^2 - \omega^2/c_0^2$. Show further that if $m^2/B^2 + n^2/C^2 < \omega^2/c_0^2\pi^2$, then

$$\phi = -\frac{a}{\mu} \cos \frac{m\pi y}{B} \cos \frac{n\pi z}{C} \sin(\omega t - \mu x),$$

where $\mu^2 = \omega^2/c_0^2 - m^2\pi^2/B^2 - n^2\pi^2/C^2$. (This problem of a *wavemaker* will be considered further in Section 4.2.)

3.3 Show that if $\phi(r, t)$ is the velocity potential for a spherically symmetric acoustic wave, where r is the polar coordinate measured from the origin, then

$$\frac{\partial^2 \phi}{\partial r^2} + \frac{2}{r}\frac{\partial \phi}{\partial r} = \frac{1}{c_0^2}\frac{\partial^2 \phi}{\partial t^2}.$$

Deduce that $r\phi$ satisfies the one-dimensional wave equation.
Acoustic waves in an infinite gas are driven by a sphere which starts oscillating at $t = 0$ so that its radius is given by $r = a(1 + \varepsilon \cos \omega t)$, where $\varepsilon \ll 1$. Show that the appropriate boundary condition for acoustic waves in $r > a$ is

$$\frac{\partial \phi}{\partial r} = -a\varepsilon\omega \sin \omega t \quad \text{on } r = a.$$

Show that for $t > 0$, the velocity potential ϕ is given by

$$\phi = \frac{1}{r}\left[\frac{a^3\varepsilon\omega^2 c^2}{c_0^2 + a^2\omega^2}\right]\left[a\omega \cos \frac{\omega}{c_0}(r - a - c_0 t)\right.$$

$$\left. + c_0 \sin \frac{\omega}{c_0}(r - a - c_0 t) - a\omega e^{(r - a - c_0 t)/a}\right]$$

for $a < r < a + c_0 t$.

R 3.4 Show that for small-amplitude waves on an incompressible stream in which
$\mathbf{u} = \nabla(Ux + \phi)$, where ϕ and the elevation η are small, the linearised
versions of the boundary conditions (3.10) and (3.11) are

$$\frac{\partial \phi}{\partial z} = \frac{\partial \eta}{\partial t} + U \frac{\partial \eta}{\partial x}$$

and

$$\frac{\partial \phi}{\partial t} + U \frac{\partial \phi}{\partial x} + g\eta = 0$$

on $z = 0$. If $\eta = a\cos(kx - \omega t)$, show that a solution of (3.8) satisfying
$\partial \phi / \partial z = 0$ on $z = -h$ and the above boundary conditions is

$$\phi = \frac{a(\omega - Uk)\cosh k(z + h)\sin(kx - \omega t)}{k \sinh kh},$$

providing $(\omega - Uk)^2 = gk \tanh kh$. Deduce that ω and k can only both be
real if $g > 0$.
Show that a solution for steady waves (with $\omega = 0$) is only possible if
$U^2 < gh$. Show also that, if $U = 0$, then, as $h \to \infty$, $\omega^2 \to gk$.

R 3.5 Show that three-dimensional Stokes waves on the surface of a running
stream of depth h can be found where the surface elevation is

$$\eta = a\cos(k_1 x + k_2 y - \omega t)$$

and the velocity potential is

$$\phi = Ux + b\cosh\sqrt{k_1^2 + k_2^2}(z + h)\sin(k_1 x + k_2 y - \omega t),$$

provided

$$(Uk_1 - \omega)^2 = g\sqrt{k_1^2 + k_2^2}\tanh\sqrt{k_1^2 + k_2^2}h$$

and

$$b = \frac{a(\omega - Uk_1)}{\sqrt{k_1^2 + k_2^2}\sinh\sqrt{k_1^2 + k_2^2}h}.$$

3.6 Small-amplitude waves propagate on the interface $z = 0$ which separates liquid
of density ρ_1 in $z > 0$ from liquid of density ρ_2 in $z < 0$. The upper liquid
is streaming with uniform velocity U in the x-direction, and the lower fluid
is at rest. If variables in the upper and lower liquid are denoted by suffices 1
and 2, respectively, and $z = \eta$ is the elevation of the interface, show that the
model (3.12)–(3.14) generalises to

$$\nabla^2 \phi_1 = 0 \text{ in } z > 0, \quad \nabla^2 \phi_2 = 0 \text{ in } z < 0,$$

with

$$\frac{\partial \phi_1}{\partial z} = \frac{\partial \eta}{\partial t} + U\frac{\partial \eta}{\partial x}, \quad \frac{\partial \phi_2}{\partial z} = \frac{\partial \eta}{\partial t}$$

and

$$\rho_1\left(\frac{\partial \phi_1}{\partial t} + U\frac{\partial \phi_1}{\partial x} + g\eta\right) = \rho_2\left(\frac{\partial \phi_2}{\partial t} + g\eta\right)$$

on $z = 0$. Show that waves for which $\eta = a\cos(kx - \omega t)$ with $k > 0$ are possible provided

$$\rho_1((\omega - Uk)^2 + gk) = \rho_2(gk - \omega^2).$$

Deduce that when $U = 0$, with $g > 0$, ω and k can only both be real if $\rho_2 \geq \rho_1$. Show also that ω and k cannot both be real when $g = 0$ and $U \neq 0$.

3.7 From (2.6) and (2.7), show that waves propagating in a vertical direction in an inhomogeneous atmosphere satisfy

$$\frac{\partial \rho}{\partial t} + \frac{\partial}{\partial z}(\rho w) = 0,$$

$$\rho\left(\frac{\partial w}{\partial t} + w\frac{\partial w}{\partial z}\right) = -\frac{\partial p}{\partial z} - g\rho$$

and

$$\left(\frac{\partial}{\partial t} + w\frac{\partial}{\partial z}\right)\left(\frac{p}{\rho^\gamma}\right) = 0.$$

Show that, in equilibrium, $\rho = \rho_s(z)$ and $p = p_s(z)$ satisfy (3.20). For acoustic waves, the variables w, $\bar{\rho} = \rho - \rho_s$ and $\bar{p} = p - p_s$ are all small. Show that the linearised equations satisfied by these variables are

$$\frac{\partial \bar{\rho}}{\partial t} + \rho_s\frac{\partial w}{\partial z} + \rho_s' w = 0,$$

$$\rho_s\frac{\partial w}{\partial t} = -\frac{\partial \bar{p}}{\partial z} - g\bar{\rho}$$

and

$$\frac{\partial \bar{p}}{\partial t} - c_s^2\frac{\partial \bar{\rho}}{\partial t} + w(p_s' - c_s^2\rho_s') = 0,$$

where $c_s^2 = \gamma p_s/\rho_s$. If we assume that gravity is negligible, show that p_s is constant and deduce that

$$\frac{\partial^2 \bar{p}}{\partial t^2} = \frac{\partial}{\partial z}\left(c_s^2\frac{\partial \bar{p}}{\partial z}\right).$$

3.8 A component of a heat exchanger consists of a uniform tube along the x-axis which contains gas and whose walls transmit heat to the gas at a rate $\partial Q(x,t)/\partial t$ per unit length. When $Q = 0$, the gas has constant speed U, density ρ_0 and pressure p_0. Show that for small heat addition, the pressure, density and velocity perturbations satisfy the equations

$$\frac{\partial \bar{\rho}}{\partial t} + \rho_0 \frac{\partial \bar{u}}{\partial x} + U \frac{\partial \bar{\rho}}{\partial x} = 0,$$

$$\rho_0 \frac{\partial \bar{u}}{\partial t} + \rho_0 U \frac{\partial \bar{u}}{\partial x} = -\frac{\partial \bar{p}}{\partial x}$$

and

$$\rho_0 c_v \left(\frac{\partial \bar{T}}{\partial t} + U \frac{\partial \bar{T}}{\partial x} \right) = \frac{p_0}{\rho_0} \left(\frac{\partial \bar{\rho}}{\partial t} + U \frac{\partial \bar{\rho}}{\partial x} \right) + \frac{\partial Q}{\partial t},$$

where $\bar{T} = \bar{p}/\rho_0 R - p_0 \bar{\rho}/\rho_0^2 R$. Deduce that

$$\left(\frac{\partial}{\partial t} + U \frac{\partial}{\partial x} \right) \left(\frac{\partial^2 \bar{\rho}}{\partial t^2} + 2U \frac{\partial \bar{\rho}}{\partial x \partial t} + (U^2 - c_0^2) \frac{\partial^2 \bar{\rho}}{\partial x^2} \right) = (\gamma - 1) \frac{\partial^3 Q}{\partial x^2 \partial t}.$$

3.9 Suppose that the Rossby Number in an incompressible rotating fluid is small so that (3.29) holds with respect to a frame rotating with angular velocity $\boldsymbol{\Omega}$. Show that if $\boldsymbol{\Omega} = (0,0,\Omega)$, then, in steady flow, $\partial \mathbf{u}/\partial z = 0$ (this is the *Taylor–Proudman Theorem*).

Now suppose that fluid fills a sealed cylindrical container which is rotating with small Rossby Number. The ends of the cylinder are flat and perpendicular to the axis of rotation save for a small finite bump on one end which protrudes into the fluid. Show that an observer rotating with the cylinder will see a two-dimensional flow perpendicular to the axis of rotation in which a "pillar" of fluid above the bump is at rest (this pillar is called a *Taylor column*).

R 3.10 A long tube containing gas at rest lies along the x-axis. In $x < 0$, the gas has density ρ_1 and sound speed c_1, while in $x > 0$, the gas has density ρ_2 and sound speed c_2. An acoustic wave described by $\phi = a \sin k(c_1 t - x)$ is incident from the region $x < 0$. Show that, at $x = 0$, $\partial \phi/\partial x$ and $\rho \partial \phi/\partial t$ are continuous, and deduce that the reflected and transmitted waves have amplitude aR and aT, respectively, where

$$R = \left| \frac{\rho_2 c_2 - \rho_1 c_1}{\rho_1 c_1 + \rho_2 c_2} \right| \quad \text{and } T = \left| \frac{2\rho_1 c_2}{\rho_2 c_2 + \rho_1 c_1} \right|.$$

(R is the *reflection coefficient* and T is the *transmission coefficient*.)

This illustrates the idea of the *impedance* of a boundary which is a generic expression used to describe the qualitative response of an inhomogeneity to an incoming wave. In this case, we can see that if $\rho_1 c_1 = \rho_2 c_2$, there is no reflected wave. Hence, when $(\rho_1 c_1 - \rho_2 c_2)$ is suitably small, we say that boundary has low impedance, while if ρ_1 or c_2 is suitably small, the transmission is weak and it has high impedance.

*3.11 In this Exercise, [ ] is used to denote the size of a discontinuous wave jump in a variable.

   (i) The vector **a** satisfies $\nabla.\mathbf{a} = 0$ and changes rapidly from one side of a surface S to the other. By integrating over a "pillbox" straddling an area Σ of S with normal **n**, and then shrinking the pillbox to zero, show that

$$[\mathbf{a}.\mathbf{n}]_-^+ = 0$$

in the limit when **a** has a jump discontinuity across S.

   (ii) The matrix $A = (A_{ij})$ satisfies the equation $\partial A_{ij}/\partial x_j = 0$. Show that if A has a jump discontinuity across S, then

$$[A\mathbf{n}]_-^+ = \mathbf{0}.$$

   (iii) Show that $(\nabla \wedge \mathbf{b})_i = -\partial A_{ij}/\partial x_j$ if $A = \begin{pmatrix} 0 & -b_3 & b_2 \\ b_3 & 0 & -b_1 \\ -b_2 & b_1 & 0 \end{pmatrix}$. Hence deduce that if $\nabla \wedge \mathbf{b} = \mathbf{0}$ and **b** has a jump discontinuity across S, then

$$[\mathbf{b} \wedge \mathbf{n}]_-^+ = \mathbf{0}.$$

   (iv) When ε and μ are spatially dependent, Maxwell's equations can be written as

$$\nabla.(\varepsilon\mathbf{E}) = 0,$$

$$\nabla.(\mu\mathbf{H}) = 0,$$

$$\mu\frac{\partial\mathbf{H}}{\partial t} = -\nabla \wedge \mathbf{E}$$

and

$$\varepsilon\frac{\partial\mathbf{E}}{\partial t} = \nabla \wedge \mathbf{H}.$$

Show that across a surface on which **E** and **H** have jump discontinuities

$$[\varepsilon\mathbf{E}.\mathbf{n}] = 0, \quad [\mu\mathbf{H}.\mathbf{n}] = 0,$$

$$[\mathbf{E} \wedge \mathbf{n}] = \mathbf{0} \text{ and } [\mathbf{H} \wedge \mathbf{n}] = \mathbf{0}.$$

Note that if conducting material is present, Maxwell's equations have to be modified to allow for current flow. Hence, these jump conditions may not be appropriate at the boundary of a conductor.

3.12 A simple model for incompressible magnetohydrodynamics is

$$\frac{\partial \mathbf{u}}{\partial t} + (\mathbf{u}.\nabla)\mathbf{u} = \frac{1}{\rho}(-\nabla p + \mathbf{j} \wedge \mathbf{B}), \quad \nabla.\mathbf{u} = 0,$$

$$\frac{\partial \mathbf{B}}{\partial t} = \nabla \wedge (\mathbf{u} \wedge \mathbf{B}), \quad \mathbf{j} = \frac{1}{\mu}\nabla \wedge \mathbf{B}, \quad \nabla.\mathbf{B} = 0,$$

where $\mathbf{u}$ is the fluid velocity, $\mathbf{j}$ is the current in the fluid and $\mathbf{B} = \mu\mathbf{H}$ (where $\mathbf{H}$ is the magnetic field and μ the magnetic permeability). The first equations are those of incompressible fluid dynamics under the influence of a magnetic body force, the remainder are Maxwell's equations for a moving medium. Show that small perturbations from the constant solution $\mathbf{u} = \mathbf{0}$, $p = p_0$, $\mathbf{B} = \mathbf{B}_0$ satisfy the wave equation with speed $|\mathbf{B}_0|\sqrt{\mu\rho}$. These waves are called *Alfvén waves*.

*3.13 Use the fact that, for any reasonable function f,

$$\int_{-\infty}^{\infty} f(x)\delta(x - y)\, dx = f(y)$$

to show that

$$\int_{-\infty}^{\infty} f(x)\delta'(x - y)\, dx = -f'(y).$$

Show further that if $f_i = n_i(x, t)\delta(u_i - \bar{u}(x, t))$, then integrating (3.53) with respect to u_i leads to (3.56). Now multiply (3.53) by u_i and integrate to get

$$\frac{\partial}{\partial t}(n_i\bar{u}) + \frac{\partial}{\partial x}(n_i\bar{u}^2) + n_i\frac{\partial \phi}{\partial x} = 0$$

and hence deduce that (3.57) holds.

*3.14 By writing $\mathbf{X} = \nabla\phi + \nabla \wedge \mathbf{\Psi}$, where $\nabla.\mathbf{\Psi} = 0$, show that the equation (3.72) for elastic waves can be satisfied if

$$\frac{\partial^2 \phi}{\partial t^2} = c_p^2 \nabla^2 \phi$$

and

$$\frac{\partial^2 \mathbf{\Psi}}{\partial t^2} = c_s^2 \nabla^2 \mathbf{\Psi},$$

where $c_p = ((\lambda + 2\mu)/\rho)^{1/2}$ and $c_s = (\mu/\rho)^{1/2}$.

Consider waves travelling in the x-direction in a semi-infinite elastic solid $z \geq 0$. Given that $\phi = \phi(x, z, t)$, $\mathbf{\Psi} = (0, -\psi(x, z, t), 0)$ and that the boundary condition on the free surface $z = 0$ is $\sigma_{i3} = 0$, where

$$\sigma_{ij} = \delta_{ij}\lambda\frac{\partial X_k}{\partial x_k} + \mu\left(\frac{\partial X_i}{\partial x_j} + \frac{\partial X_j}{\partial x_i}\right),$$

show that on $z = 0$

$$\lambda \left(\frac{\partial^2 \phi}{\partial x^2} + \frac{\partial^2 \phi}{\partial z^2} \right) + 2\mu \left(\frac{\partial^2 \phi}{\partial z^2} - \frac{\partial^2 \psi}{\partial x \partial z} \right) = 0$$

and

$$2\frac{\partial^2 \phi}{\partial x \partial z} - \frac{\partial^2 \psi}{\partial x^2} + \frac{\partial^2 \psi}{\partial z^2} = 0.$$

*3.15 Starting from (3.82) and (3.83), show that for a material that has traversed the elastic/plastic boundary and is deformed uniaxially,

$$e_1^p + 2e_2^p = 0.$$

Show further that

$$\frac{\partial e_1^e}{\partial a} = \frac{\partial e_2^e}{\partial a} = \frac{1}{2}\frac{\partial e_1^p}{\partial a}$$

and hence that

$$\frac{\partial^2 X}{\partial a^2} = 3\frac{\partial e_1^e}{\partial a}.$$

Finally, show that

$$\frac{\partial \tau_1}{\partial a} = (3\lambda + 2\mu)\frac{\partial e_1^e}{\partial a}$$

and deduce the wave equation (3.86).

*3.16 A *hyperelastic* material is defined in [28] to be one in which the stresses depend on the gradient of a strain energy function $W(F_{ij}^e)$. For a uniaxial deformation as defined in Section 3.8.4.2, this leads to the constitutive equations

$$\tau_1 = \frac{\rho}{\rho_0} F_1^e \frac{\partial W}{\partial F_1^e}, \quad \tau_2 = \frac{\rho}{2\rho_0} F_2^e \frac{\partial W}{\partial F_2^e}.$$

Now assume that the mechanical response is elastically linear so that

$$W = \frac{\lambda}{2}(F_1^e + 2F_2^e - 3)^2 + \mu \left((F_1^e - 1)^2 + 2(F_2^e - 1)^2 \right).$$

Show that

$$\tau_1 = \frac{\rho F_1^e}{\rho_0} \left((\lambda + 2\mu)F_1^e + 2\lambda F_2^e - (3\lambda + 2\mu) \right)$$

and

$$\tau_2 = \frac{\rho F_2^e}{\rho_0} \left(\lambda F_1^e + 2(\lambda + \mu)F_2^e - (3\lambda + 2\mu) \right)$$

and that these expressions reduce to (3.78) and (3.79) for small strains. Now assume that the applied stress is so large that σ_Y can be neglected. Show that in this limit

$$F_1^e = F_2^e = \left(\frac{\rho_0}{\rho}\right)^{1/3}$$

and deduce that the material behaves like a barotropic fluid with pressure

$$(3\lambda + 2\mu)\left(\frac{\rho}{\rho_0}\right)\left(\left(\frac{\rho}{\rho_0}\right)^{1/3} - 1\right).$$

*3.17 A sound source with frequency ω moves along a tube with speed V for time $t > 0$. You are given that velocity potential satisfies

$$\frac{\partial^2 \phi}{\partial x^2} = \frac{1}{c^2}\frac{\partial^2 \phi}{\partial t^2}, \quad x \neq Vt,$$

where ϕ is continuous at the source $x = Vt$, and the velocity jump across $x = Vt$ is $\cos \omega t$. If the gas is initially at rest, show that, if $V < c$,

$$\frac{2\omega c}{c^2 - V^2}\phi = \begin{cases} -\sin\dfrac{(ct - x)\omega}{c - V}, & Vt < x < ct \\[3mm] -\sin\dfrac{(ct + x)\omega}{c + V}, & -ct < x < Vt. \end{cases}$$

Deduce that the Doppler frequency shift between observers ahead of and behind the source is

$$\omega c\left(\frac{1}{c - V} - \frac{1}{c + V}\right) = \frac{2\omega c V}{c^2 - V^2}.$$

Chapter 4
Theories for Linear Waves

Looking back at the models derived in the last chapter, we see that many of them comprise linear partial differential equations in time and at least one space variable, together with linear boundary conditions. Moreover, in most of the equations, many of the terms have constant coefficients. We therefore start this chapter by reviewing the mathematical methodologies that are available for the analysis of such models.

4.1 Wave equations and Hyperbolicity

It is well known that systems of linear partial differential equations can be classified into a hierarchy which has "hyperbolic" models at one end and "elliptic" models at the other [43]. Hyperbolic models are probably the best understood and for such a system the *Cauchy problem*, in which appropriate data is prescribed at some initial time is, in general, a well-posed problem. Moreover, much is known about how the solution depends on the data via "regions of influence" and "domains of dependence" in the space of the independent variables. These concepts depend crucially on the fact that the *characteristics* (in two dimensions) or the *characteristic manifolds* (in three or more dimensions) are real for a hyperbolic system. However, this information does not necessarily give us any detailed knowledge of the solution itself, let alone an explicit analytic solution.

The situation can be illustrated with reference to the wave equation (3.4) in one space dimension,

$$\frac{\partial^2 \phi}{\partial x^2} = \frac{1}{c^2}\frac{\partial^2 \phi}{\partial t^2};\tag{4.1}$$

for most of this chapter, we drop the suffix zero from c_0 for convenience. Since it is known (see [43]) that the quasilinear partial differential equation

$$A\frac{\partial^2 \phi}{\partial x^2} + 2B\frac{\partial^2 \phi}{\partial x \partial t} + C\frac{\partial^2 \phi}{\partial t^2} = D,\tag{4.2}$$

© Springer Science+Business Media New York 2015
H. Ockendon, J.R. Ockendon, *Waves and Compressible Flow*,
Texts in Applied Mathematics 47, DOI 10.1007/978-1-4939-3381-5_4

where A, B, C depend only on $x, t, \phi, \partial\phi/\partial x$ and $\partial\phi/\partial t$, is hyperbolic if $B^2 > AC$, and its characteristics are given by

$$C\left(\frac{dx}{dt}\right)^2 - 2B\frac{dx}{dt} + A = 0,$$

we can see at once that (4.1) is hyperbolic and its characteristics are the lines $x \pm ct = $ constant. Moreover, if *Cauchy data*

$$\phi(x,0) = f(x) \quad \text{and} \quad \frac{\partial\phi}{\partial t}(x,0) = g(x) \tag{4.3}$$

are prescribed at $t = 0$ for all x, where, additionally, f, g are only non-zero in a finite interval $a < x < b$, then it is only possible for the solution to be non-zero in $a-ct < x < b+ct$. This is the *region of influence* of the interval (a, b). Even without the general theory, these results follow directly from the *D'Alembert solution*

$$\phi = \frac{1}{2}[f(x - ct) + f(x + ct)] + \frac{1}{2c}\int_{x-ct}^{x+ct} g(s)\, ds. \tag{4.4}$$

Yet another way to look at the solution of (4.1) is by "factorising" the differential operators and writing the equation in the form

$$\left(\frac{\partial}{\partial x} \mp \frac{1}{c}\frac{\partial}{\partial t}\right)\left(\frac{\partial\phi}{\partial x} \pm \frac{1}{c}\frac{\partial\phi}{\partial t}\right) = 0,$$

so that it follows that $\partial\phi/\partial x \pm (1/c)(\partial\phi/\partial t)$ are constant on the lines $x \pm ct = $ constant. These quantities are called *Riemann Invariants*, and a further integration shows that the general solution is $\phi = F(x-ct) + G(x+ct)$ for arbitrary functions F and G.

We note that by considering Cauchy data with "compact support" (i.e. data that is only non-zero on a finite interval of the x-axis), we have found solutions that are not analytic everywhere; hence, we have run the risk of ending up with a solution which is not differentiable enough for (4.1) to make sense. Indeed (4.1) and (4.3) can apparently become an ill-posed problem when we consider, for example,

$$f(x) = \begin{cases} 1, & x > 0 \\ \\ 0, & x < 0 \end{cases}, \quad g(x) = 0 \,.$$

Then, away from $x + \lambda t = 0$, the function $\phi = f(x + \lambda t)$ satisfies both (4.1) and (4.3) for any constant λ. We will return to the discussion of such singular solutions in Chapter 6 but will not let this interrupt the discussion for the moment.

It is sad but ineluctable that (4.1) is almost the only model from Chapter 3 whose general solution can be written down explicitly, as in (4.4). One other such case is that of sound waves with spherical symmetry when (3.6) reduces to

$$\frac{\partial^2 \phi}{\partial r^2} + \frac{2}{r}\frac{\partial \phi}{\partial r} = \frac{1}{c^2}\frac{\partial^2 \phi}{\partial t^2}, \tag{4.5}$$

where r is the spherical polar coordinate. At first sight, this looks worse than (4.1) because the extra term does not have a constant coefficient, but writing $r\phi = \Phi$ leads to (4.1) for Φ, so that the general solution is

$$\phi = \frac{1}{r}[F(r - ct) + G(r + ct)]. \tag{4.6}$$

We thus see that acoustic waves in one dimension and in three dimensions are closely related. But things are not so simple in two dimensions. Writing (3.6) in cylindrical polar coordinates and assuming that the flow is axisymmetric leads to

$$\frac{\partial^2 \phi}{\partial r^2} + \frac{1}{r}\frac{\partial \phi}{\partial r} = \frac{1}{c^2}\frac{\partial^2 \phi}{\partial t^2};$$

there are non-radial solutions of the form $[(F(r-ct)+G(r+ct))r^{-1/2}]\cos\theta/2$, but the term $(1/r)(\partial\phi/\partial r)$ really does make life harder. We will return to this equation and to the fascinating question of how waves depend on the dimensionality of the space within which they are propagating in Section 4.9.

"Wave equations" such as (3.6), (3.25) and (3.32) act as a marvellous springboard for a mathematical treatment of wave motion and might provide a basis for the statement that "wave equations are hyperbolic equations". However, surface gravity waves are described by an *elliptic* partial differential equation, so how is it that the boundary conditions can allow wave solutions? Also how can we reconcile hyperbolicity with the observation that if we seek acoustic waves that vary harmonically in time in three dimensions by writing $\phi = \mathrm{Rl}\,(\Phi(x, y, z)e^{-i\omega t})$, then we are left with the *elliptic* equation $\nabla^2 \Phi + (\omega^2/c^2)\Phi = 0$?

These questions suggest that we need a more general idea than that of hyperbolicity if we are to encompass many of the waves that occur in nature.

4.2 Fourier Series, Eigenvalues and Resonance

Fourier analysis is one of the most powerful methods for the analysis of linear equations, especially ones which have constant coefficients. Irrespective of whether the partial differential equation is hyperbolic, elliptic or parabolic, such an equation will have solutions that can be obtained by the method of separation of variables, which leads to solutions that are products of exponential functions (with either real or imaginary argument).[1] By summing solutions of this type, it is possible to generate quite general explicit solutions which are often more convenient even than an exact representation such as D'Alembert solution (4.4) of the one-dimensional wave equation.

---

[1] Separation of variables can sometimes be applied to variable-coefficient equations, as will be seen later.

If we consider acoustic waves or waves on a finite string satisfying (4.1) and with boundary conditions $\phi = 0$ at $x = 0, L$, it is easy to separate the variables and appeal to the theory of Fourier series in order to obtain

$$\phi = \sum_{n=1}^{\infty}(a_n \cos \frac{n\pi ct}{L} + b_n \sin \frac{n\pi ct}{L}) \sin \frac{n\pi x}{L}. \tag{4.7}$$

For this finite domain, this form of solution is much easier to deal with than the D'Alembert solution (4.4) which will involve an infinite series of reflecting waves. For a Cauchy problem, where the initial values are given by (4.3), the coefficients a_n and b_n can easily be found to be

$$a_n = \frac{2}{L} \int_0^L f(x) \sin \frac{n\pi x}{L} dx \tag{4.8a}$$

and

$$b_n = \frac{2}{n\pi c} \int_0^L g(x) \sin \frac{n\pi x}{L} dx. \tag{4.8b}$$

This analysis requires that the functions f and g satisfy certain smoothness conditions, and the series (4.7) will only converge at points where ϕ is continuous.

The solution (4.7) for one-dimensional waves in a closed container, such as an organ pipe, consists of a sum of "eigenmodes" or "normal modes" which can be thought of as the infinite-dimensional generalisation of the normal modes encountered when considering small oscillations in classical mechanics. These oscillations are perpetual motions (assuming there is no damping) which can exist in the absence of any long-term forcing; they just need to be initiated by some given non-zero initial conditions. If the wave is forced by persistent non-zero boundary conditions at $x = 0, L$, we can still use this form of the solution but we will need to add in a "particular solution" of (4.1) which satisfies the forcing condition and then the calculation for a_n and b_n will be different. For example, suppose that we wish to solve (4.1) given

$$\phi = 0 \text{ on } x = 0 \text{ and } \phi = \cos \omega t \text{ on } x = L,$$

in addition to the usual initial conditions (4.3). Then we can write the solution as

$$\phi = \frac{\sin \omega x/c \cos \omega t}{\sin \omega L/c} + \sum_{n=1}^{\infty}\left(a_n \cos \frac{n\pi ct}{L} + b_n \sin \frac{n\pi ct}{L}\right) \sin \frac{n\pi x}{L} \tag{4.9}$$

as long as $\omega L/c\pi$ is not in integer, and then, applying the initial conditions, we obtain

$$a_n = \frac{2}{L} \int_0^L (f(x) - \frac{\sin \omega x/c}{\sin \omega L/c}) \sin \frac{n\pi x}{L} dx,$$

and b_n is as given in (4.8b).

We make the important remark that the Fourier series representation (4.7) is not chosen simply because Fourier series are a convenient and familiar way of representing mathematical functions on a finite interval; the form of the "modes" defined by the terms in (4.7) are a direct result of separating the variables in (4.1) and applying the zero (homogeneous) boundary conditions at $x = 0$ and L. If non-trigonometric eigenfunctions had emerged as a result of separation of variables, it would have been appropriate to employ a generalised Fourier series expansion in which these eigenfunctions were used as a basis (see Exercise 4.2). This procedure is often needed for the construction of normal modes in more general wave models.

Normal modes, whether in terms of trigonometric functions or not, are of great practical importance because of the phenomenon of *resonance*. We met this idea in Exercise 3.2 and a second example is the revelation in (4.9) that the periodic forced solution will not exist if $\omega = n\pi c/L$ for some integer n. In crude terms, resonance is the surprisingly large amplitude response that occurs when the boundary forcing is at one of the "natural frequencies" (or "normal frequencies") of the unforced system.

We now consider how the idea of Fourier analysis can be used to help us to understand resonance in a more general situation. Intuitively, we expect a decomposition into eigenmodes to be possible for any linear undamped unforced wave model in a finite domain. To see this mathematically, we represent the wave model by

$$\mathcal{L}\phi = \frac{1}{c^2}\frac{\partial^2 \phi}{\partial t^2} \qquad \text{in } D, \tag{4.10}$$

with

$$\phi = 0 \qquad \text{on } \partial D,$$

where $\mathcal{L}$ is some linear spatial elliptic operator. We then seek waves of frequency ω by writing[2]

$$\phi = \text{Rl}\,(\Phi e^{-i\omega t}). \tag{4.11}$$

This leads to the problem

$$\mathcal{L}\Phi + \frac{\omega^2}{c^2}\Phi = 0 \ \text{ in } D, \tag{4.12}$$

with

$$\Phi = 0 \ \text{ on } \partial D,$$

which is an eigenvalue problem where the eigenvalues ω^2/c^2 determine the natural frequencies of the system (4.10). There is an enormous literature on such problems and the dependence of the eigenvalues on the operator $\mathcal{L}$ and the geometry of D. We will not discuss this further here except to say that under "nice" conditions, and when there is no damping, the possible values of ω will be discrete, real and positive in any closed *resonator* or finite domain D. Typically, the discrete numbers ω will "grow linearly" so that if, say, in a one-dimensional problem, the eigenvalues ω_n are arranged in increasing order of magnitude, then $\omega_n = O(n)$ as $n \to \infty$. In particular, for the one-dimensional acoustic oscillator in $0 < x < L$, we see from (4.7) that $\omega_n = n\pi c/L$ and, for the three-dimensional version for waves in a rectangular box with sides of length A, B and C (Exercise 3.2),

---

[2] The negative sign is taken in the exponent for reasons that will become apparent later.

$$\omega_{lmn} = \pi c \left[\frac{l^2}{A^2} + \frac{m^2}{B^2} + \frac{n^2}{C^2} \right]^{1/2}, \tag{4.13}$$

where l, m, n are all integers.

Armed with this information about eigenvalues, we can quickly encapsulate the phenomenon of resonance in mathematical terms. Suppose that an acoustic resonator is forced to oscillate by having part or all of its boundary moving with frequency ω. If we seek a periodic solution to (3.6) in the form (4.11), we find that

$$\nabla^2 \Phi + \frac{\omega^2}{c^2} \Phi = 0, \tag{4.14}$$

with Φ prescribed and non-zero on the boundary ∂D. By the Fredholm Alternative (see [43]), we can assert that this problem will have no solution whenever ω/c is one of the eigenvalues of the solution of (4.14) with zero boundary conditions. Thus, the phenomenon of resonance is simply a reflection of the Fredholm Alternative.

From a different viewpoint, if the resonator is at rest, and we start to drive it periodically at one of the eigenfrequencies, we find that the response grows linearly with time, thereby destroying any possibility of an eventual periodic response. In practice, this unlimited growth is usually mitigated by the effects of some damping in the system, as will be illustrated in several of the examples in Section 4.4.1. However, in certain cases, the amplitude of the response may become so large as to invalidate the assumptions that were built into the linear approximation. Then, as we will see in Chapter 5, we have to reconsider the perturbation schemes we used in Chapter 3 so as to bring nonlinear terms into play.

We have only defined normal modes and resonances when the wave motion is confined in a closed container. Quite a different situation applies when the motion occurs in a region that extends to infinity in all directions. If we excite such a system by a *transient* localised forcing, we expect all the energy to propagate to infinity, as in the solution (4.6), and the motion will eventually die away even if there is no damping. If, however, a *periodic* forcing is maintained, then we will again be able to look for solutions of the form $\phi = \mathrm{Rl}\,(e^{-i\omega t}\Phi)$ with Φ satisfying (4.14). In this case, the conditions to be imposed at infinity are less obvious and we will return to such problems in Section 4.5.1.

An interesting configuration that is halfway between a bounded "interior" problem and an infinite "exterior" problem is found in a *waveguide*. This is a device in which waves are directed to propagate in a semi-infinite channel; the reflection from the walls of the channel allows the propagation to be unattenuated, unlike the spherically symmetric wave given by (4.6). We can understand this most easily by solving the problem for two-dimensional acoustic waves in a channel with fixed walls at $y = 0, b$. The velocity potential ϕ will satisfy

$$\frac{\partial^2 \phi}{\partial x^2} + \frac{\partial^2 \phi}{\partial y^2} = \frac{1}{c^2} \frac{\partial^2 \phi}{\partial t^2},$$

with $\partial \phi / \partial y = 0$ on $y = 0, b$, so writing $\phi = \mathrm{Rl}\,(\Phi e^{-i\omega t})$ now leads to solutions of the form

$$\Phi = \cos \frac{n\pi y}{b}(Ae^{ikx} + Be^{-ikx}), \tag{4.15}$$

where $k^2 = \omega^2/c^2 - n^2\pi^2/b^2$ and n is any integer.[3] From this, we can immediately discern the crucial attribute of many wave guides; this is the fact that a harmonic wave can only propagate in the x-direction if ω exceeds the so-called *cut-off* frequency $c\pi/b$; if this is not the case, k will not be real and (4.15) will not represent a propagating wave (see Exercise 3.2).

We conclude this section with one piece of jargon. So far all our linear wave models have been posed as *evolution problems* in which time appears as an independent variable and we are thus in the "time domain". However, the representations in this section in terms of eigenmodes have inevitably led us to equations like (4.12) or (4.14) in which the only independent variables are spatial, and it is assumed that all time variations are harmonic with frequency ω. Such problems are said to be posed in the *frequency domain* and will be considered in more detail in Section 4.5. However, before going down this route we turn our attention to the Fourier representation of problems in infinite domains.

4.3 Fourier Integrals and the Method of Stationary Phase

In order to represent solutions of models for waves propagating in infinite domains, we start by indicating how the theory for Fourier series can be extended to apply to nonperiodic functions. This leads us to *Fourier transform* theory.

Noting that a general, sufficiently smooth, $2l$ periodic function $f(x)$ has the Fourier series

$$f(x) = \frac{a_0}{2} + \sum_{1}^{\infty} a_n \cos \frac{n\pi x}{l} + b_n \sin \frac{n\pi x}{l}, \tag{4.16}$$

where

$$a_n + ib_n = \frac{1}{l}\int_{-l}^{l} f(x)\left(\cos \frac{n\pi x}{l} + i\sin \frac{n\pi x}{l}\right) dx,$$

we can see at once that an alternative formulation of (4.16) is

$$f(x) = \sum_{-\infty}^{\infty} c_n e^{-in\pi x/l}, \quad c_n = \frac{1}{2l}\int_{-l}^{l} f(x)e^{in\pi x/l}\, dx,$$

where $c_n = \frac{1}{2}(a_n + ib_n)$ and $a_n = a_{-n}$, $b_n = -b_{-n}$. This gives us the vital clue as to how to deal with *nonperiodic* functions. All we need to do formally is to put $n\pi/l = k$, and let $l \to \infty$, to obtain the Fourier integral transform formulae

---

[3] The reason for the minus sign in (4.11) is now apparent, because it means the first term in (4.15) represents waves propagating in the positive x-direction.

$$\bar{f}(k) = \int_{-\infty}^{\infty} f(x)e^{ikx}\,dx \tag{4.17}$$

and

$$f(x) = \frac{1}{2\pi} \int_{-\infty}^{\infty} \bar{f}(k)e^{-ikx}\,dk. \tag{4.18}$$

Of course this leaves open important questions of convergence which we will not address in general terms; when these questions loom large, we will deal with them on a case-by-case basis.

A trivial illustration of this method is the solution of the one-dimensional wave equation (4.1) on $-\infty < x < \infty$. By multiplying (4.1) by e^{ikx} and integrating from $x = -\infty$ to $x = \infty$, we see at once that the Fourier transform of ϕ is $\bar{F}(k)e^{ikct} + \bar{G}(k)e^{-ikct}$, where $\bar{F}$ and $\bar{G}$ are general functions, and hence

$$\phi = \frac{1}{2\pi} \int_{-\infty}^{\infty} \bar{F}(k)e^{-ik(x-ct)}\,dk + \frac{1}{2\pi} \int_{-\infty}^{\infty} \bar{G}(k)e^{-ik(x+ct)}\,dk.$$

This is a solution containing waves of all wave number k which can be seen to be the sum of two travelling waves

$$\phi = F(x - ct) + G(x + ct),$$

as obtained earlier. In wave propagation, the Fourier transform variable k is usually called the *wavenumber*, and solutions of (4.1) that are proportional to $e^{\pm ikx}$ have wavelength $2\pi/k$.

The solution of the general initial value problem for two-dimensional surface gravity waves is less trivial. Suppose that we consider waves on water of infinite depth which is initially at rest with a surface elevation $z = \eta_0(x)$. Then, we can take the Fourier transform of equations (3.12), (3.13) and (3.14) by writing

$$\bar{\phi}(k, z, t) = \int_{-\infty}^{\infty} \phi(x, z, t)e^{ikx}\,dx,$$

$$\bar{\eta}(k, t) = \int_{-\infty}^{\infty} \eta(x, t)e^{ikx}\,dx,$$

and, assuming that $|\phi|$ and $|\nabla\phi| \to 0$ as $|x| \to \infty$, we get

$$\frac{\partial^2 \bar{\phi}}{\partial z^2} - k^2 \bar{\phi} = 0,$$

with

$$\frac{\partial \bar{\phi}}{\partial z} = \frac{\partial \bar{\eta}}{\partial t}, \quad \frac{\partial \bar{\phi}}{\partial t} + g\bar{\eta} = 0 \quad \text{on } z = 0.$$

In order for $\bar{\phi}$ to decay as $z \to -\infty$, we are forced to choose

$$\bar{\phi} = \bar{A}(k, t)e^{|k|z},$$

where $|k|\bar{A} = \partial\bar{\eta}/\partial t$, $\partial\bar{A}/\partial t = -g\bar{\eta}$, so that $\bar{A}$ satisfies

$$\frac{\partial^2 \bar{A}}{\partial t^2} = -g|k|\bar{A}.$$

Then, since $\bar{A} = 0$, $d\bar{A}/dt = -g\bar{\eta}_0$ at $t = 0$, we find that

$$\bar{A} = -\bar{\eta}_0 \sqrt{g/|k|} \sin \sqrt{g|k|}t \text{ and } \bar{\eta} = \bar{\eta}_0 \cos \sqrt{g|k|}t,$$

so that

$$\eta(x,t) = \frac{1}{2\pi} \int_{-\infty}^{\infty} \bar{\eta}_0(k) \cos \sqrt{g|k|}t \, e^{-ikx} \, dk. \tag{4.19}$$

It is unfortunate that the integrand in this integral is, as it stands, a non-analytic function of k, but we recall from complex variable theory that we can define $|k| = \lim_{\varepsilon \to 0} \sqrt{k^2 + \varepsilon^2}$ as long as we define the "branch" of the function $\sqrt{k^2 + \varepsilon^2}$ correctly. This gives us a *theoretical* solution of the general gravity wave problem in two dimensions, but of what use is it? Only for a very few functions $\bar{\eta}_0$ will we be able to integrate (4.19) explicitly, and, from the numerical point of view, (4.19) is a superposition of harmonic waves which is difficult to represent accurately with a computer, especially if $\bar{\eta}_0(k)$ is appreciable for large values of k. Nevertheless, there is an ingenious asymptotic method for determining how (4.19) behaves when x and t are large, and this is the last piece of fundamental methodology to be described in this section.

It is well known that integrals involving exponentials with large arguments can be evaluated asymptotically by *Laplace's Method* and this theory is described by Hinch [26]. To illustrate the idea by an example, suppose we want to evaluate

$$I_1 = \int_{-\infty}^{\infty} A(k)e^{xg(k)} \, dk$$

asymptotically as $x \to \infty$, where $g(k)$ is a real function that takes its largest value at $k = k_0$. Then we find that, when $g''(k_0) \neq 0$,

$$I_1 \sim \frac{A(k_0)}{\sqrt{2\pi x |g''(k_0)|}} e^{xg(k_0)} \left(1 + O\left(\frac{1}{x}\right)\right).$$

However, the modification of this argument to consider an integral like (4.19) makes things more difficult. Nonetheless, if we consider

$$I_2 = \int_{-\infty}^{\infty} A(k)e^{ixg(k)} \, dk,$$

where g is still real valued, we can still assert that the main contribution to the integral as $x \to \infty$ comes from values of k for which $g'(k) = 0$. This results from the key observation that the real and imaginary parts of the integrand oscillate with a wavelength of $O(x^{-1})$ except at these critical values of k, where the wavelength is $O(x^{-1/2})$. The sketch in Figure 4.1 illustrates how, say, the real part of the integrand behaves. We can see that away from points where $g'(k) = 0$, the contribution to the integral from neighbouring values of k will cancel out much more efficiently than it will in the neighbourhood of such points. Thus, if $g(k)$ has one turning point at $k = k_0$, it can be shown that

$$I_2 \sim \frac{A(k_0)e^{ixg(k_0)}}{\sqrt{x}} \int_{-\infty}^{\infty} e^{ig''(k_0)s^2/2}\, ds.$$

Using contour integral methods, we can show that, if $\lambda > 0$,

$$\int_{-\infty}^{\infty} e^{i\lambda s^2}\, ds = (1+i)\sqrt{\frac{\pi}{2\lambda}},$$

and so

$$\int_{-\infty}^{\infty} A(k)e^{ixg(k)}\, dk \sim \frac{\sqrt{\pi}A(k_0)e^{ixg(k_0)}(1+i)}{\sqrt{g''(k_0)x}} \qquad (4.20)$$

as $x \to \infty$, if $g''(k_0) > 0$. This method is called the *method of stationary phase*, and the formula can easily be adapted to deal with large negative values of x or with cases where $g''(k_0) < 0$. We must, however, issue the warning that while it is quite easy to find the $O(1/x)$ correction to the lowest order approximation by using Laplace's method, it is often impossible to find the next term explicitly when using the method of stationary phase.

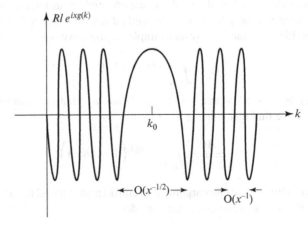

Fig. 4.1 $Rl\, e^{ixg(k)}$ near a turning point of $g(k)$.

The estimate (4.20) immediately allows us to make predictions from (4.19). In order to see what happens for large x, t, we imagine an observer moving with constant speed V by writing $x = Vt$, and then we let $x, t \to \infty$ keeping V constant. From (4.19) we are led to consider integrals of the form

$$\int_{-\infty}^{\infty} \bar{\eta}_0(k) e^{i(-kV \pm \sqrt{g|k|})t} \, dk.$$

The exponents $-kV \pm \sqrt{g|k|}$ have turning points at $k = \pm g/4V^2$ and we can see straightaway that, after a long time, the dominant waves seen by the observer have one or other of these wave numbers. Using formula (4.20), we can see that, as $t \to \infty$,

$$\eta(Vt, t) \sim \text{Rl} \left\{ \frac{C_1 e^{igt/4V} + C_2 e^{-igt/4V}}{\sqrt{t}} \right\}, \tag{4.21}$$

for some complex constants C_i; the terms in C_1 and C_2 correspond to waves travelling in the positive and negative x-direction, with $k = g/4V^2$, respectively. Thus, as V increases, the frequency and wave number of the dominant observed waves decreases and their wavelength increases. Alternatively, a stationary observer far from the initial disturbance will see waves of gradually increasing frequency and wave number as time goes by. These predictions can be verified for the special case in which η_0 is localised[4] near $x = 0$ (see Exercise 4.4).

More generally, we notice that, in the far field, waves of wave number k are only observed to dominate by an observer travelling at speed V if $|V| = \frac{1}{2}\sqrt{g/k}$, and this speed is called the *group velocity* of waves of wave number k. This idea can be generalised to models where (4.19) is replaced by

$$\eta(x, t) = \frac{1}{2\pi} \int_{-\infty}^{\infty} \bar{\eta}_0(k) \cos \omega(k)t \, e^{-ikx} \, dk.$$

Then the dominant contribution to the integral occurs when k is such that

$$V = \frac{d\omega}{dk}$$

and this is the mathematical definition of group velocity for such a wave model.

This is all dramatically different to the type of solution one gets when solving the acoustic wave equation (4.1) with localised initial data. In that case $\omega(k) = \pm ck$, for any value of k, and the disturbance is eventually propagated at constant speed c without change of shape, so that results like (4.21) are not observed. The phenomenon we have just described for gravity waves is a manifestation of the *dispersive* nature of the system; dispersion occurs whenever the speed of waves varies with the wave number, and we will consider this idea in more generality in the next section.

---

[4] They can also be observed by throwing a stone into a large deep pond.

4.4 Dispersion and Group Velocity

We have already remarked that most of the models in Chapter 3 admit separable solutions that are products of exponential functions in space and time. Indeed, if we write the solution in each case as

$$\phi = \text{Rl}\,[A e^{i\mathbf{k}.\mathbf{x}-i\omega t}], \tag{4.22}$$

where, as usual, ω is the frequency, $\mathbf{k}$ is a vector in the direction of the travelling wave and $|\mathbf{k}|$ is the wave number, then we almost always find that ω satisfies a *dispersion relation* of the form

$$\omega = \omega(\mathbf{k}).$$

We have already seen explicit examples of such relations in Exercises 3.2, 3.4, 3.5 and 3.6 and, in Section 4.3, we used the fact that, for waves on deep water, $\omega = \sqrt{g|k|}$. In this case, we observed that the non-constancy of the group velocity $d\omega/dk$ led to the phenomenon of dispersion.

   We anticipate the fact that the stationary phase argument used in Section 4.3 can be applied to multiple Fourier integrals of the form

$$\iiint A(\mathbf{k}) e^{i(\mathbf{k}.\mathbf{x}-\omega t)}\, d\mathbf{k},$$

and will predict that the dominant contribution seen by an observer at $\mathbf{x} = \mathbf{V}t$ for large t comes from the values of $\mathbf{k}$ for which

$$\nabla_{\mathbf{k}}(\mathbf{k}.\mathbf{V} - \omega(\mathbf{k})) = \mathbf{0}, \tag{4.23}$$

where $\nabla_{\mathbf{k}} = (\partial/\partial k_1, \partial/\partial k_2, \partial/\partial k_3)$ and $\mathbf{k} = (k_1, k_2, k_3)$. From (4.23) we see that

$$\mathbf{V} = \nabla_{\mathbf{k}}\omega, \tag{4.24}$$

and this motivates us to define $\nabla_{\mathbf{k}}\omega$ as the *group velocity* of the waves with wave number $\mathbf{k}$. The *phase velocity* of these waves is

$$\frac{\omega \mathbf{k}}{|\mathbf{k}|^2}, \tag{4.25}$$

and (4.24),(4.25) are only equal if $\omega = c|\mathbf{k}|$, where c is a constant. In all other circumstances the wave speed varies with $\mathbf{k}$ and the system is dispersive. Also, we will soon see that several of the models lead to ω becoming complex when $\mathbf{k}$ is real and the possible implications of this will be discussed in Section 4.4.2.

   We now look at the models derived in Chapter 3 in the light of these ideas.

4.4.1 Dispersion Relations

Collecting together the dispersion relations for the systems considered in Chapter 3, we will now see that a number of these systems are dispersive. Most of the following results emerge trivially from substituting solutions of the form (4.22) into the relevant field equations without imposing any boundary conditions.

(i) *Acoustic Waves.*

Equation (3.4) leads to the dispersion relation

$$\omega^2 = c^2|\mathbf{k}|^2,$$

so these waves are always non-dispersive.

However, we note that this dispersion relation is only true when there are no boundaries. We recall that, in Section 4.2, we showed that in a waveguide, where we solved the acoustic wave equation subject to boundary conditions on $y = 0$ and b, we were led to (4.15) and the dispersion relation

$$\omega^2 = c^2\left(k^2 + \frac{n^2\pi^2}{b^2}\right), \tag{4.26}$$

where n is a positive integer. The group velocity will be

$$V = \frac{ck}{(k^2 + n^2\pi^2/b^2)^{1/2}},$$

and thus waves in an acoustic waveguide are dispersive.

(ii) *Surface Gravity Waves*

Two-dimensional waves satisfy (3.12) with boundary condition (3.13) and (3.14) on water of depth h provided ϕ is proportional to $\mathrm{Rl}\cosh k_3\,(z + h)e^{i(k_1x + k_2y - \omega t)}$. This gives

$$\omega^2 = g|\mathbf{k}|\tanh|\mathbf{k}|h, \tag{4.27}$$

where $|\mathbf{k}|^2 = k_1^2 + k_2^2$, and these waves are dispersive. The group velocity is given by

$$V = \frac{g}{2\omega|\mathbf{k}|}(\tanh|\mathbf{k}|h + |\mathbf{k}|h\,\mathrm{sech}^2|\mathbf{k}|h)\mathbf{k}.$$

Note that, as $h \to 0$, (4.27) reduces to

$$\omega^2 = gh|\mathbf{k}|^2,$$

so that waves on shallow water are non-dispersive and their phase speed is $\sqrt{gh}$. However, when $h \to \infty$, we see that

$$\omega^2 = g|\mathbf{k}|$$

so that deep water waves are dispersive.

(iii) *Waves on a Stratified Fluid*

A solution of type (4.22) only works for these waves if N is constant in (3.26). This can happen if ρ_0 is an exponential function of z and in this case

$$\omega^2 = \frac{N^2(k_1^2 + k_2^2)}{|\mathbf{k}|^2 + ik_3 N^2 g^{-1}}, \tag{4.28}$$

which is clearly dispersive. Furthermore, ω will be complex when $\mathbf{k}$ is real and we discuss the implications of this in the next section.

(iv) *Waves in a Rotating Fluid*

For waves in a rotating fluid governed by (3.32), the dispersion relation is

$$\omega^2 = \frac{4\Omega^2 k_3^2}{|\mathbf{k}|^2}, \tag{4.29}$$

and these waves are dispersive with group velocity

$$\mathbf{V} = \frac{2\Omega}{|\mathbf{k}|^3}(-k_1 k_3, -k_2 k_3, k_1^2 + k_2^2).$$

Such waves are sometimes called *inertial waves*.

(v) **Waves in Dissociating Gases*

Equation (3.37) leads to the dispersion relation

$$\overline{\tau}\omega^3 + i\omega^2 - c_{f_0}^2 \overline{\tau}k_1^2\omega - ic_{e_0}^2 k_1^2 = 0. \tag{4.30}$$

This equation shows how gas dynamics can easily lead to complicated dispersion relations. Here, ω is the root of a complex cubic equation and there will always be dispersion. Also when k_1 is real, we can see that ω must be complex.

(vi) **Waves in Viscous Incompressible Flow*

There are no interesting unperturbed flows $\psi_0(y)$ for which the Orr–Sommerfeld equation (3.42) can be solved explicitly. However, all solutions are linear combinations of functions of the form $\tilde{\psi} = \mathrm{RL}\,[g(y)e^{i(kx - \omega t)}]$ where, on substituting into (3.42), we find that $g(y)$ satisfies

$$\nu\left(\frac{d^2}{dy^2} - k^2\right)^2 g = ik\left[\left(\psi_0'(y) - \frac{\omega}{k}\right)\left(\frac{d^2}{dy^2} - k^2\right)g - \psi_0'''g\right]. \tag{4.31}$$

Together with suitable homogeneous boundary conditions, (4.31) is an eigenvalue problem which determines the relation between ω and k in principle. Unfortunately, this dispersion relation, which is inevitably complex, can only be solved for ω numerically.

(vii) **Electromagnetic Waves*

In a vacuum, (3.45) implies that

$$\omega^2 = c^2|\mathbf{k}|^2,$$

so that we appear to have no dispersion, as for acoustic waves. However, if we calculate the dispersion relation directly from (3.43) and (3.44), which is a sixth-order system, we find

$$\omega^2(\omega^2 - c^2|\mathbf{k}|^2)^2 = 0.$$

The root $\omega = 0$ can be rejected since the corresponding eigensolution is not in general consistent with the equation $\nabla.\mathbf{H} = \nabla.\mathbf{E} = 0$, but the fact that the dispersion relation is really a degenerate quartic has implications for the Fourier analysis mentioned above. Also, this degeneracy is related to the polarisation effect which was mentioned in Section 3.8.3.

In a conducting medium, using (3.47) and (3.48) in place of (3.43), the dispersion relation is

$$\omega^2 - c^2|\mathbf{k}|^2 + i\sigma\mu c^2\omega = 0, \tag{4.32}$$

so that conduction causes dispersion as well as making the dispersion relation complex.

(viii) *Acoustic Waves in Plasmas*

For the equation (3.61), the dispersion relation for cold ions is

$$\omega^2 = \frac{k^2}{1 + k^2}, \tag{4.33}$$

which means that ion-acoustic waves in a plasma travel without dispersion for small wave numbers but at large wave numbers *all* waves have frequency unity. In this latter limit, both the phase velocity and the group velocity tend to zero. However, for the cold-electron model (3.65)–(3.67), the dispersion relation reduces to $\omega^2 = 1$ for all values of k. In this situation, the value $\omega = 1$ is called the *plasma frequency* and any spatial waveform will oscillate at this single frequency with zero group velocity and a phase speed which is inversely proportional to k.

The system (3.68)–(3.70) is much more subtle than any wave previously encountered. This is largely because it is the first model we have seen that incorporates randomness. This also means that it is more convenient to obtain the dispersion relation by taking a Laplace transform in time rather than directly substituting a plane wave into the underlying model. When we do this, assuming some arbitrary initial data, the dispersion relation appears in the denominator of the Laplace transform and, if the initial data is zero, the dispersion relation is the condition for the existence of a non-zero Laplace transform.

We begin by seeking time-dependent solutions of (3.68)–(3.70) in which

$$\tilde{f} = \mathrm{Rl}\,\hat{f}(u_e, t; k)e^{ikx},$$

$$\phi = \mathrm{Rl}\,\hat{\phi}(t; k)e^{ikx}$$

and

$$\tilde{n} = \mathrm{Rl}\,\hat{n}(t;k)e^{ikx},$$

so that

$$\frac{\partial \hat{f}}{\partial t} + iu_e k\hat{f} + ikf_0'\hat{\phi} = 0$$

and

$$-k^2\hat{\phi} = \hat{n} = \int_{-\infty}^{\infty} \hat{f}\,du.$$

Hence, if $\tilde{f} = 0$ at $t = 0$,

$$\hat{f} = \frac{if_0'(u_e)}{k} \int_0^t \hat{n}(\tau)e^{iku_e(\tau-t)}\,d\tau$$

and integrating over u_e leads to

$$\hat{n} = \frac{i}{k} \int_{-\infty}^{\infty} f_0'(u) \left(\int_0^t \hat{n}(\tau)e^{iku(\tau-t)}\,d\tau \right) du. \tag{4.34}$$

If we define the Laplace transform[5]

$$\bar{n}(p;k) = \int_0^{\infty} \hat{n}(t;k)e^{-pt}dt$$

and apply it to (4.34), we find that

$$\bar{n}(p;k) = \frac{i}{k}\bar{n}(p;k) \int_{-\infty}^{\infty} \frac{f_0'(u)\,du}{p + iku}, \tag{4.35}$$

where we need to take $\mathrm{Rl}\,p > 0$ so that the Laplace transform and the integral in (4.35) are well defined. Writing $p = -i\omega$, the dispersion relation for non-zero $\bar{n}$ is thus

$$\frac{1}{k^2} \int_{-\infty}^{\infty} \frac{f_0'(u)\,du}{u - \omega/k} = 1, \tag{4.36}$$

where $\mathrm{Im}\,\omega > 0$. This unusual dispersion relation is easy to understand in the special case of cold electrons with zero drift velocity when $f_0(u) = \delta(u)$ and then (4.36) reduces to $\omega^2 = 1$ as expected.

The restriction to values of ω such that $\mathrm{Im}\,\omega > 0$ appears at first sight to limit its value, but it is possible to circumvent this difficulty by using the delicate mathematical process of analytic continuation in the complex ω-plane. This can be done explicitly in the special case, where

---

[5] This transform can be derived formally from the Fourier transform (4.17) as described in [43].

$$f_0(u) = \frac{\varepsilon}{\pi(u^2 + \varepsilon^2)}$$

even though this is not a realistic distribution function. When $\varepsilon \downarrow 0, f_0(u) \to \delta(u)$ but, for $\varepsilon > 0$, we need to find the zeros of the function

$$F(\omega) = \frac{1}{k^2} \int_{-\infty}^{\infty} \frac{f_0'(u)\, du}{u - \omega/k} - 1$$

$$= \frac{-2\varepsilon}{k^2\pi} \int_{-\infty}^{\infty} \frac{u\, du}{(u - i\varepsilon)^2 (u + i\varepsilon)^2 (u - \omega/k)} - 1.$$

First, we consider the case when $k > 0$ so that $\text{Im}\,(\omega/k) > 0$, and the only singularity of the integrand in the lower half of the complex u-plane is at $u = -i\varepsilon$. Then, by considering the integral round a large semicircle in $\text{Im}\, u < 0$ and using the calculus of residues we find that

$$F(\omega) = \frac{1}{(\omega + i\varepsilon k)^2} - 1.$$

If we now put $F(\omega) = 0$ to satisfy (4.36), we find that $\text{Im}\,\omega < 0$, violating the assumption made in (4.36). To get around this problem we observe that $F(\omega)$ is a holomorphic function which can therefore be analytically continued from $\text{Im}\,\omega > 0$ into the whole ω-plane apart from the double pole at $\omega = -i\varepsilon k$. Now we can solve $F(\omega) = 0$ and see that ω satisfies the complex dispersion relation

$$\omega = \pm 1 - i\varepsilon k \quad \text{if} \quad k > 0. \tag{4.37}$$

When $k < 0$, a similar argument can be applied to find that

$$F(\omega) = \frac{1}{(\omega - i\varepsilon k)^2} - 1$$

and so

$$\omega = \pm 1 + i\varepsilon k \quad \text{if} \quad k < 0. \tag{4.38}$$

A similar calculation can be carried out for the more realistic case when f_0, given by (3.71), is the Maxwellian distribution

$$f_0(u) = \frac{1}{\sqrt{\pi\varepsilon}} e^{-u^2/\varepsilon}.$$

In this case, $\text{Im}\,\omega$ again turns out to be negative but it is exponentially small in ε.

(ix) *Elastic Waves and Plastic Waves*
Elastic waves are governed by (3.72) and if we write

$$\mathbf{X} = \text{Rl}\,(\mathbf{A}e^{i\mathbf{k}.\mathbf{x} - i\omega t}),$$

we find that the dispersion relation for longitudinal waves, where $\mathbf{A}$ is parallel to $\mathbf{k}$, is

$$\omega^2 = \left(\frac{\lambda + 2\mu}{\rho}\right) |\mathbf{k}|^2, \tag{4.39}$$

while for transverse waves, where $\mathbf{A}$ is perpendicular to $\mathbf{k}$,

$$\omega^2 = \frac{\mu}{\rho}|\mathbf{k}|^2. \tag{4.40}$$

Both these waves are non-dispersive, but there is the possibility of "mode conversion" if energy is transferred from longitudinal waves to transverse waves or vice versa. This is an important phenomenon in seismic waves as shown in Exercises 4.7 and 4.11.

Linear plastic waves are modelled by (3.86), and the dispersion relation is

$$\omega^2 = \frac{3\lambda + 2\mu}{3\rho_0}|\mathbf{k}|^2 \tag{4.41}$$

which is again non-dispersive.

4.4.2 Stability and Dissipation

The dispersion relations listed above can be broadly classified into ones for which ω is necessarily real for all real $\mathbf{k}$ and those in which ω is complex for at least some real $\mathbf{k}$.

In the former case, all waves will be superpositions of harmonic waves proportional to $\exp[i(\mathbf{k}.\mathbf{x} - \omega t)]$ which propagate through all space without spatial or temporal decay as $|\mathbf{x}|, t \to \infty$. However, we have seen that the presence of boundaries can have a dramatic effect on propagating waves.

When ω is complex for at least some values of real $\mathbf{k}$, the large time behaviour will depend crucially on the sign of $\operatorname{Im}\omega$. If $\operatorname{Im}\omega < 0$ for all real $\mathbf{k}$, as is the case in (4.37) and (4.38), then all waves decay in time and the mathematical model is said to be *linearly stable*. The prototype situation for which this occurs is in the theory of diffusion or heat conduction when the model is

$$\frac{\partial^2 u}{\partial x^2} = \frac{\partial u}{\partial t}$$

and this leads to the dispersion relation $\omega = -ik^2$. Models for which $\operatorname{Im}\omega$ is always negative are called *dissipative* or *damped* and this is exemplified in the examples above by electromagnetic waves in a conducting medium when the dispersion relation is (4.32). This means that electromagnetic waves in a conductor are always damped and in particular, when σ is large, time-harmonic electromagnetic waves can only be sustained in a thin layer near the surface of the conductor. This is called the "skin effect" for good conductors.

A particularly subtle form of damping is revealed by the dispersion relation (4.36) for electron-acoustic waves in plasmas. When the ions are cold, these waves can propagate with any wavelength but at a single frequency. However, the randomness introduced by a small, finite ion temperature leads to a damping that is exponentially small for realistic density distributions. Such damping is called *Landau damping* [39].

On the other hand, if, for some real **k**, Im ω is positive, then waves will, in general, grow without bound as $t \to \infty$ and the model is linearly unstable. This can happen in several of the examples in Section 4.4.1, one of the most dramatic being when we change the sign of g in (4.27), so that the heavy fluid is above the lighter air. Then the positivity of Im ω leads to the simplest example of the famous *Rayleigh–Taylor* instability. This is a particularly catastrophic instability because ω is imaginary for all k and is large for small wavelengths $2\pi/k$. As shown in Exercise 3.6, a very similar situation is revealed if we seek waves on the interface between two inviscid fluids moving parallel to each other and to the surface but with different velocities; such a surface is called a *vortex sheet*. In this case, the instability is called the *Kelvin–Helmholtz* instability. Both these instabilities can be stabilised by the introduction of surface tension at the interface; surface tension is a powerful mechanism at short wave lengths and allows the existence of capillary waves as described, for example, in Drazin and Reid [18] (see also Exercise 4.6).

Linear instability also occurs for the generalised wave equation (3.37). From the corresponding dispersion relation (4.30), we see that for a wave travelling with speed $c = \omega/k_1$,

$$\omega = \frac{i(c^2 - c_{eo}^2)}{c_{fo}^2 - c^2}.$$

Hence, if c lies between c_{e_o} and c_{f_0}, Im $\omega > 0$ and the waves are linearly unstable. This means that for small times the first term in (3.37) dominates and waves propagate with speed c_{f_0}, but eventually these waves evolve into ones travelling with speed c_{e_0}.

Finally we note that, although the dispersion relation corresponding to differential equation (4.31) can usually only be found numerically, we can gain some insight for channel flow $u = U(1 - y^2/L^2)$ in the case when the Reynolds number UL/ν is large. By scaling ψ_0 and g with UL, y with L, k with $1/L$ and ω with U/L and putting $\psi_0(y) = y - \frac{1}{3}y^3$, (4.31) becomes

$$\frac{1}{Re}\left(\frac{d^2}{dy} - k^2\right)^2 g = ik\left(1 - y^2 - \frac{\omega}{k}\right)(g'' - k^2 g) + 2ikg. \tag{4.42}$$

If we let Re $\to \infty$ and also assume $g = g' = 0$ at the walls $y = \pm 1$, it emerges from numerical calculations that Im $\omega > 0$ when $k = 1$. Such a solution of (4.42) is called a *Tollmien–Schlichting wave*, even though it is unstable, and is believed to be important in the early stages of turbulence.

4.4.3 *Other Approaches to Group Velocity

Up to now, we have used large-time asymptotics as the motivation for introducing group velocity, but several other approaches are possible.

From the physical viewpoint, group velocity can be interpreted as the velocity with which the mean energy in a particular mode is transported; this idea is described in Lighthill [37]. A more elementary theoretical motivation comes from the following simple example, which gives an intuitive idea of the difference between phase and group velocity. We consider two sinusoidal wave trains of equal amplitude but slightly different wave number and frequency. The sum of these waves is

$$a\sin(k_1 x - \omega_1 t) + a\sin(k_2 x - \omega_2 t) = 2a\cos(\Delta k x - \Delta \omega t)\sin(kx - \omega t),$$

where $k_1 = k + \Delta k, k_2 = k - \Delta k, \omega_1 = \omega + \Delta \omega, \omega_2 = \omega - \Delta \omega$. Thus if Δk and $\Delta \omega$ are small, the result is a slowly modulated wave of amplitude $2a\cos(\Delta k x - \Delta \omega t)$, as shown in Figure 4.2. We can see that the so-called envelope of the wave crests and troughs travels with speed $\Delta \omega / \Delta k$ which is approximately the group velocity of these waves.

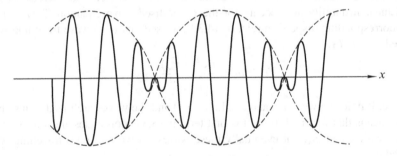

Fig. 4.2 Superposition of two harmonic waves

Alternatively, we can take the asymptotic approach as in Section 4.3, but now we will use it directly on the equations rather than on the Fourier transform solution. This means we need to use the *WKB expansion method* to determine the long time or far field solution of the linear wave model. To do this, we have to introduce a large artificial parameter ε^{-1} to represent the length and time scales and we rescale the independent variables by $\mathbf{x} = \varepsilon^{-1}\mathbf{X}$ and $t = \varepsilon^{-1}T$. Then we write

$$\phi = \text{Rl}\left(\hat{A}(\mathbf{X}, T)e^{iu(\mathbf{X},T)/\varepsilon}\right),$$

to lowest order in ε as $\varepsilon \to 0$. We note that we can do this only after the introduction of ε and taking the limit $\varepsilon \to 0$; otherwise the functions u (the phase) and $\hat{A}$ (the amplitude) would not be uniquely defined[6] for a given ϕ. We will return to this idea in more detail in Section 4.8.

Now we can see that, since ε is small,

$$\frac{\partial \phi}{\partial T} \sim \text{Rl} \left[\left(i \frac{\hat{A}}{\varepsilon} \frac{\partial u}{\partial T} \right) e^{iu/\varepsilon} \right],$$

and

$$\frac{\partial^2 \phi}{\partial T^2} \sim \text{Rl} \left[-\frac{\hat{A}}{\varepsilon^2} \left(\frac{\partial u}{\partial T} \right)^2 e^{iu/\varepsilon} \right],$$

to lowest order in ε. Now when ϕ satisfies (4.22),

$$\frac{\partial \phi}{\partial t} = \text{Rl} \left[-i\omega A e^{i\mathbf{k}.\mathbf{x}-i\omega t} \right] \quad \text{and} \quad \frac{\partial^2 \phi}{\partial t^2} = \text{Rl} \left[-\omega^2 A e^{i\mathbf{k}.\mathbf{x}-i\omega t} \right],$$

and so we can identify $\partial u/\partial T$ with $-\omega$ and $\hat{A}$ with A. Similarly, $\bar{\nabla} u$ can be identified with $\mathbf{k}$ in (4.22), where $\bar{\nabla} = (\partial/\partial X_1, \partial/\partial X_2, \partial/\partial X_3)$. Hence, if the dispersion relation is

$$\omega = \omega(\mathbf{k}),$$

we can immediately infer that u will satisfy the equation

$$\frac{\partial u}{\partial T} + \omega(\bar{\nabla} u) = 0, \tag{4.43}$$

which is a first-order equation for the phase of the far field solution. It is also possible to find an equation for the amplitude $\hat{A}$ (see Section 4.8.1). Even for non-dispersive systems, (4.43) will be a *nonlinear* partial differential equation; it only reduces to a linear equation in the case of one space dimension when $\omega = ck$.

Reverting to our identification of ω with $-\partial u/\partial T$ and $\mathbf{k}$ with $\bar{\nabla} u$, but now regarding ω and $\mathbf{k}$ as dependent variables with arguments $\mathbf{X}, T$, we see from (4.43) that

$$-\bar{\nabla}\omega = \bar{\nabla} \frac{\partial u}{\partial T} = \frac{\partial}{\partial T}(\bar{\nabla} u) = \frac{\partial \mathbf{k}}{\partial T},$$

or[7]

$$\frac{\partial \mathbf{k}}{\partial T} + \sum_{i=1}^{3} \frac{\partial \omega}{\partial k_i} \bar{\nabla} k_i = \mathbf{0}. \tag{4.44}$$

---

[6] Note that with a conventional power series asymptotic expansion $\phi \sim \sum_0^\infty \varepsilon^n \phi_n$, we can define the terms recursively via $\phi_0 = \lim_{\varepsilon \to 0} \phi$, $\phi_1 = \lim_{\varepsilon \to 0} (\phi - \phi_0)/\varepsilon$, etc. Equally for an expansion $\phi \sim e^{u/\varepsilon} \sum_{n=0}^\infty A_n \varepsilon^n$, where u is real, we can define $u = \lim_{\varepsilon \to 0} \varepsilon \log \phi$. However, in this case with u real, no such definition is possible because u now measures the oscillations in ϕ rather than its magnitude. The best that can be done is to say that $|\nabla u|^2 = \lim_{\varepsilon \to 0}(-\varepsilon^2 \nabla^2 \phi/\phi)$.

[7] Note that equation (4.44) can be thought of as "conservation of wave number" if $\omega \mathbf{k}$ is interpreted as "wave number flux".

We can most easily analyse this equation in one dimension, when (4.44) reduces to

$$\frac{\partial k}{\partial T} + \frac{d\omega}{dk}\frac{\partial k}{\partial X} = 0,$$

which, since ω is only a function of k, has the general solution

$$k = F\left(X - \frac{d\omega}{dk}T\right)$$

for an arbitrary function F. Thus, k and ω are constant along the lines

$$X - \frac{d\omega}{dk}T = \text{constant},$$

and so, for large times, we again see that the waves with wave number k will dominate if we travel with the group velocity of the waves. We can also say that $\partial u/\partial T$ and $\bar{\nabla}u$ will be constant on these lines and so $u/\varepsilon \simeq kx - \omega t$. All this can be generalised to three dimensions and confirms, in a more general setting, the results in Section 4.3 for surface waves but, unfortunately, no obvious rules have emerged concerning the relation between the magnitude and direction of the phase and group velocity.

We will now put these general considerations on one side and consider some concrete situations with boundaries in the frequency domain. It turns out to be convenient to distinguish between problems that are stationary ($\omega = 0$) from those that are not and we will start by considering the latter.

4.5 The Frequency Domain

4.5.1 Homogeneous Media

All the phenomena to be discussed in this section can be illustrated with reference to the simplest *acoustic model* for monochromatic waves of prescribed frequency ω. Then, writing $\phi = \text{Rl}\,(\Phi(\mathbf{x})e^{-i\omega t})$ as usual, the equation for Φ is (4.14) which we write as

$$(\nabla^2 + k^2)\Phi = 0, \tag{4.45}$$

where $k = \omega/c$. This is *Helmholtz' equation*, which is an elliptic partial differential equation that appears to be a simple generalisation of Laplace's equation. However, the mathematics associated with (4.45) turns out to be very different from the usual theories for Laplace's equation.

The first fundamental classification that must be made of waves in the frequency domain is the distinction between *interior* and *exterior* problems.

In the former case, we expect to have a boundary condition prescribed for Φ or $\partial\Phi/\partial n$, or some combination thereof, everywhere on a closed boundary, and we are confronted with the problem of the solvability of (4.45) in D with, say, the boundary condition

$$\alpha\Phi + \beta\frac{\partial\Phi}{\partial n} = f \quad \text{on } \partial D.$$

If f, the forcing term, is identically zero and α and β have the same sign, we have a classical eigenvalue problem for those values $k = k_i$ for which a nontrivial solution Φ_i exists. A great deal is known about such problems and much of it is described in Courant and Hilbert [14]. If, however, the forcing term f is non-zero, then we can use the eigenfunctions of the unforced problem to establish an integrability condition. This is another application of the Fredholm Alternative mentioned in Section 4.2. Suppose, for example, that

$$\nabla^2\Phi_i + k_i^2\Phi_i = 0 \quad \text{in } D \text{ and } \Phi_i = 0 \text{ on } \partial D$$

and

$$\nabla^2\Phi + k^2\Phi = 0 \quad \text{in } D \text{ with } \Phi = f \text{ on } \partial D.$$

From Green's Second Theorem, which states that

$$\int_D (\Phi\nabla^2\Phi_i - \Phi_i\nabla^2\Phi)\, dV = \int_{\partial D} \left(\Phi\frac{\partial\Phi_i}{\partial n} - \Phi_i\frac{\partial\Phi}{\partial n}\right) dS,$$

we see that if $k = k_i$, the solution Φ can only exist if

$$\int_{\partial D} f\frac{\partial\Phi_i}{\partial n}\, dS = 0.$$

If this condition is satisfied, then the solution for Φ will not be unique since it can only be determined to within a term $\lambda\Phi_i$, where λ is any constant. This is simply a more general way of stating the problem of resonance since $\omega_i = ck_i$ are the natural frequencies of the system.

Things are quite different when we examine the *exterior problem* in which boundary conditions are still imposed on ∂D, but will not be sufficient by themselves to determine the physically relevant solution. Suppose, for example, we consider the problem of an oscillating sphere of radius a so that Φ is defined in $r > a$. The boundary condition

$$\phi = \text{Rl}\,(e^{-i\omega t}) \quad \text{on } r = a,$$

implies that $\Phi = 1$ on $r = a$. Then, solving (4.45) with spherical symmetry leads to the solution

$$\Phi = \frac{A}{r}e^{ik(r-a)} + \frac{B}{r}e^{-ik(r-a)}, \tag{4.46}$$

where $A + B = a$. Even if we impose boundedness on Φ as $r \to \infty$, we are still short of a second equation relating A and B. What we have to remember is that, by posing the problem in the frequency domain, we are considering a sphere that has been oscillating for a very long time so that the system has settled down to a solution periodic in time. But we have implicitly assumed that, while this motion was being set up, no other sources of waves have been sending disturbances *towards* the sphere from anywhere in $r > a$. Thus, the waves described by (4.46) can only *radiate outwards* from $r = a$ and this means that the second term in (4.46), which represents an *inward* travelling wave in the time domain, is not physically relevant.[8] Thus, the solution is given by taking $A = a$ and $B = 0$, and the argument which has led to this conclusion is referred to as "imposing a *radiation condition* at infinity".

This example suggests the hypothesis, which can be proved, that exterior frequency domain problems are well posed (that is to say the solution exists, is unique and depends continuously on the given data) as long as we impose the *Sommerfeld radiation condition*

$$r\left(\frac{\partial \Phi}{\partial r} - ik\Phi\right) \to 0 \qquad (4.47)$$

as $r \to \infty$. The pre-multiplier r is used in (4.47) because we know from (4.46) that $r\Phi$ is $O(1)$ as $r \to \infty$. The corresponding far field behaviour in one and two dimensions dictates the radiation conditions

$$\frac{\partial \Phi}{\partial x} - ik\Phi \to 0 \text{ as } x \to \infty$$

and

$$r^{1/2}\left(\frac{\partial \Phi}{\partial r} - ik\Phi\right) \to 0 \text{ as } r \to \infty,$$

respectively (where r is now a two-dimensional polar coordinate), and we will return to this dependence on dimensionality in Section 4.9 (see also Exercise 4.9).

This analysis of problems in which waves are radiating to infinity from a finite oscillator or "radiator" can, in principle, be extended to what are perhaps the most important frequency domain problems where incoming waves are *scattered* by a finite obstacle. Such problems arise naturally in applications ranging from oil exploration to radar and from harbour design to tomography.

4.5.2 Scattering Problems in Homogeneous Media

We suppose that a finite obstacle with boundary ∂D is "irradiated" or "insonified"[9] by a plane wave travelling in the positive x-direction and given by $\Phi = Ae^{ikx}$. We take the boundary condition on ∂D to be $\Phi = 0$, which corresponds to a "hard" reflector, but we could equally well model a "soft" reflector by $\partial \Phi / \partial n = 0$. To solve this problem, all we need to do is to consider $\hat{\Phi} = \Phi - Ae^{ikx}$ and solve Helmholtz' equation (4.45) for $\hat{\Phi}$ with $\hat{\Phi} = -Ae^{ikx}$ on ∂D and the appropriate Sommerfeld condition at infinity. This is another problem with a vast literature and we will mention briefly three aspects, each illustrated by a "canonical" problem.

---

[8] Note that had ϕ been written as $\text{Rl}\,(\Phi e^{i\omega t})$, the first term would have been unacceptable.

[9] The term used depends on whether we are referring to light waves or sound waves.

(i) Reflection

Suppose that the body in the above problem is smooth and two-dimensional. Locally, near any point on the body, the solution to (4.45) will consist of a combination of Fourier modes or plane waves so that

$$\Phi = \sum_{l,m} A_{lm} e^{ilx+imy},$$

where $l^2 + m^2 = k^2$ and we use l, m here rather than k_1, k_2 to avoid "suffix clutter". Suppose we choose the tangent to the body at the point $(0,0)$ along the y-axis with the x-axis along the inward normal. Then it is clear that for every incoming wave of the form $A_{lm} e^{i(lx+my)}$ there must be a *reflected wave* of the form

$$-A_{lm} e^{i(-lx+my)}$$

if the boundary condition on $x = 0$ is to be satisfied.[10] Hence, every plane wave whose constant phase lines, or *wavefronts*, are $lx+my =$ constant induces a reflected plane wave whose constant phase lines are $lx-my =$ constant. For Helmholtz' equation, the normals to the wavefronts are called the *rays* and hence we have *specular reflection* in which the angle of incidence of the incoming ray equals the angle of reflection of the outgoing ray, as shown in Figure 4.3.

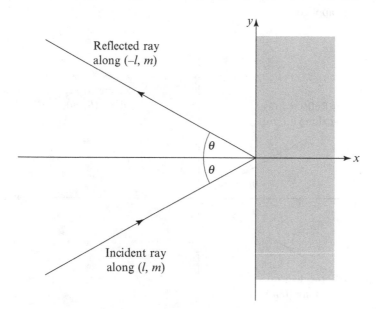

Fig. 4.3 Wave reflections at a plane boundary.

---

[10] The amplitude of the reflected wave would be different if the boundary condition was of the form $\alpha \Phi + \beta (\partial \Phi / \partial n) = 0$.

(ii) Refraction

The phenomenon of refraction occurs when waves in a medium in which the wave speed is c_0 irradiate a body that is capable of transmitting waves through its interior, where the ambient wave speed is c_1. Again, thinking of the wave $A_{lm}e^{i(lx+my)}$ impinging on the surface $x = 0$, we will now have a reflected wave given by $Be^{i(-lx+my)}$ in $x < 0$ and a transmitted wave $Ce^{i(l_1x+m_1y)}$ in $x > 0$. Here $l_1^2 + m_1^2 = \omega^2/c_1^2 = (c_0^2/c_1^2)k^2$ and, no matter what other continuity conditions we impose on $x = 0$, we must have "continuity of wave number" so that $m_1 = m$. If θ_r is the angle of refraction and θ_i is the angle of incidence, we see from Figure 4.4 that

$$\tan \theta_i = \frac{m}{l} = \frac{m}{\sqrt{k^2 - m^2}}, \quad \sin \theta_i = \frac{m}{k},$$

and

$$\tan \theta_r = \frac{m}{l_1} = \frac{m}{\sqrt{(c_0^2/c_1^2)k^2 - m^2}}, \quad \sin \theta_r = \frac{c_1 m}{c_0 k},$$

and so

$$\sin \theta_r = \frac{c_1}{c_0} \sin \theta_i, \qquad (4.48)$$

which is *Snell's law of refraction*. This condition leads to the possibility of *total internal reflection* if $c_1/c_0 \sin \theta_i > 1$ (see Exercise 4.6). The amplitudes of the reflected and refracted waves will depend on the exact form of the boundary conditions that are applied.

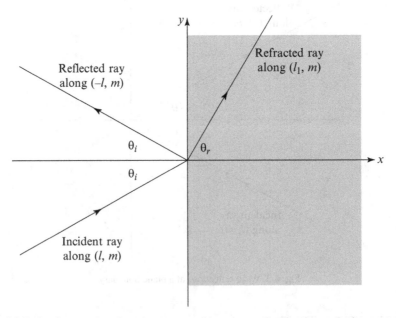

Fig. 4.4 Refraction at an interface: the wave speed is c_0 in $x < 0$ and c_1 in $x > 0$ where $c_1 > c_0$.

The phenomena of reflection and refraction apply locally to all scatterers that are smooth enough to have a tangent plane to which the irradiation is not tangential. Hence they give good intuition as to the way in which many scatterers respond to incoming waves. Alas, non-smooth scatterers often radiate most intensely, which is explained by the next phenomenon.

(iii) Diffraction
Diffraction is a much more complicated phenomenon that cannot be described simply in terms of a few plane waves as was possible for reflection and refraction. It occurs when a plane wave is incident on an obstacle in a situation where *infinitely many* new plane waves are generated. This can happen for instance when the obstacle is not smooth or when the incident wave is tangential to the body.

It is possible, by using ingenious asymptotics, to solve both of these problems approximately when k is large and to write down the intensity of the diffracted field at infinity, and examples of the directions of the scattered field are indicated schematically on the plots in Figures 4.13 and 4.14. In each case, a radiation condition has to be applied to the solution after the incident field has been subtracted out. A better idea of the scattering that can be expected will emerge from the discussion in Section 4.8 on high-frequency waves.

We remark that these three scenarios only apply when there is just one wave speed at any point of the medium. When waves can travel at more than one speed, as is the case for linear elasticity (Section 3.8.3), then the phenomenon of *mode conversion* can occur; as shown in Exercise 4.11, a wave with speed c_p incident on a rigid boundary will reflect as a combination of waves moving with speed c_p and c_s.

4.6 Inhomogeneous Media

One especially interesting phenomenon is the propagation of waves through materials whose properties vary in space, an elementary example being refraction. A one-dimensional analysis of the propagation of waves through a periodic medium reveals some of the effects that can occur. We consider a conceptual generalisation of (3.4) in the form

$$\frac{\partial^2 \phi}{\partial x^2} = P(x)\frac{\partial^2 \phi}{\partial t^2},$$

where $P = 1/c_0^2(x)$ is a positive function with period 2π in x. This equation could, for example, represent waves on a string of variable density or waves passing normally through a variable elastic medium such as seismic waves in stratified rock.

4.6.1 The Frequency Domain

In the frequency domain the problem becomes

$$\frac{d^2\Phi}{dx^2} + \omega^2 P(x)\Phi = 0 \tag{4.49}$$

which, when P is periodic, is *Hill's equation*. If $P = a + b\cos x$, the equation (4.49) becomes the *Mathieu equation* which has been much studied (Arscott [3]). It can be shown (see Exercise 4.13) that solutions of the Mathieu equation are rarely periodic and that, depending on the value of ω, solutions will either (i) grow or decay as $x \to \infty$ or (ii) be quasi-periodic.[11] The types of solutions possible for different values of the parameters are illustrated in Figure 4.5 where waves can propagate only for values of ω in the shaded "pass bands", where the solution is quasi-periodic. The key feature is that, in the so-called stop bands, which are the complements of the pass bands in Figure 4.5, the medium causes waves to decay exponentially if they try to propagate in the x-direction. Such exponential decay (or growth) would of course also occur if k^2 in (4.45) were complex or even negative. Thus, we can make an analogy between behaviour in stop bands and complex "effective" values of k. In such cases, we lose intuition and can no longer make analogies with acoustic waves in homogeneous media where c^2 is inherently positive. For instance, in the situation illustrated in Figure 4.4, suppose an electromagnetic wave rather than an acoustic wave is incident on a cleverly designed periodic medium such as a photonic crystal and the frequency is such that the effective value of ω^2/c_1^2 is complex, then it is not clear which branch of $(c_1^2)^{1/2}$ is relevant in determining the refracted wave. This leads to the possibility that "negative refraction"[12] can occur in which θ_r becomes negative.

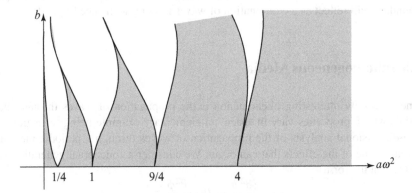

Fig. 4.5 Solutions of the Mathieu equation for small b and $a > 0$.

It can sometimes be instructive to model a one-dimensional smoothly varying inhomogeneous medium as a composition of parallel homogeneous layers and use the ideas of Section 4.5.2 at the interface between each layer. We will just look at

---

[11] A quasi-periodic solution consists of a sum of periodic terms with non-commensurate periods.

[12] In optics, it is traditional to work with the refractive index which is proportional to $1/c$ and is negative when θ_r is negative.

one layer, the analysis of which is important in the transmission and reflection of waves through thin coatings. Here, we only consider acoustic waves propagating in the x-direction through a fluid of density ρ_1 and consider the effect of introducing a layer of density ρ_2 in $0 < x < h$. Concerning the transition from medium 1 to medium 2 at $x = 0$, if the potentials in the frequency domain are Φ_1 in $x < 0$ and Φ_2 in $x > 0$, the conditions to be satisfied at $x = 0$ are

$$\rho_1 \Phi_1 = \rho_2 \Phi_2 \quad \text{and} \quad \frac{d\Phi_1}{dx} = \frac{d\Phi_2}{dx},$$

which represent the continuity of pressure and velocity at the interface. If the incident wave is given by

$$\Phi_1 = e^{ik_1 x},$$

we could proceed directly by solving Helmholtz' equation in the three regions $x < 0, 0 < x < h, h < x$, with appropriate boundary conditions, as in Exercise 3.10. However, here we use an alternative approach in which the waves are almost thought of as "particles". To do this, we note that the incident wave will engender a reflected wave $R_{12} e^{-ik_1 x}$ in $x < 0$ and a transmitted wave $T_{12} e^{ik_2 x}$ in $h > x > 0$, where

$$R_{12} = \frac{Z_2 - Z_1}{Z_2 + Z_1}, \quad T_{12} = \frac{\rho_1}{\rho_2} \left(\frac{2Z_2}{Z_1 + Z_2} \right),$$

and where $Z_i = \rho_i / k_i$ is called the *wave impedance* (see Exercise 3.10). Similarly, the reflection and transmission coefficients for a wave going from medium 2 to medium 1 are

$$R_{21} = \frac{Z_1 - Z_2}{Z_1 + Z_2} = -R_{12} = -R,$$

say, and

$$T_{21} = \frac{\rho_2}{\rho_1} \left(\frac{2Z_1}{Z_1 + Z_2} \right) = \frac{\rho_2}{\rho_1} (1 - R).$$

Now if the incident wave $\Phi_1 = e^{ik_1 x}$ impinges on the layer, it will suffer repeated reflections and transmissions as follows:

(i) The reflected wave $Re^{-ik_1 x}$ from the boundary $x = 0$.
(ii) The transmitted wave $T_{12} e^{ik_2 x}$ from $x = 0$ which will reflect at $x = h$ as $T_{12} e^{ik_2 h} R_{21} e^{-ik_2(x-h)}$ and eventually emerge into $x < 0$ as the "one-bounce" reflected wave $T_{12} e^{2ik_2 h} R_{21} T_{21} e^{-ik_1 x}$.
(iii) The "two-bounce" reflected wave $T_{12} e^{4ik_2 h} R_{21}^3 T_{21} e^{-ik_1 x}$ and so on. Thus, the total reflected wave in $x < 0$ will be

$$\left\{ R - T_{12} T_{21} R e^{2ik_2 h} \left(\sum_{n=0}^{\infty} R^{2n} e^{2ink_2 h} \right) \right\} e^{-ik_1 x},$$

which can be simplified to

$$\frac{R(1 - e^{2ik_2h})}{(1 - R^2 e^{2ik_2h})} e^{-ik_1x}.$$

Thus we can see that the layer does not reflect at all if $k_2h = n\pi$, where n is an integer; when $k_2h = \pi$, this is called a "half-wavelength" non-reflecting layer.

4.6.2 *The Time Domain

We conclude this section by describing the application of the theory of homogenisation (also known as the method of multiple scales) to unidirectional wave propagation perpendicular to the layers of a periodic medium, and in particular to one in which the width of each layer is much smaller than a typical wavelength. In such a situation, the wave speed can be written as $c(x/\varepsilon)$, where ε is a small parameter. The simplest paradigm model for this situation is

$$\frac{\partial \phi}{\partial t} + c\left(\frac{x}{\varepsilon}\right)\frac{\partial \phi}{\partial x} = 0, \tag{4.50}$$

where x varies by $O(1)$ and c is a given periodic function with period ε in x. The solution of this equation is

$$\phi = f\left(\int_0^x \frac{ds}{c(s/\varepsilon)} - t\right),$$

where f is an arbitrary function. It is easily shown that

$$\int_0^x \frac{ds}{c(s/\varepsilon)} = x\left(\overline{\frac{1}{c}}\right) + \varepsilon p\left(\frac{x}{\varepsilon}\right), \tag{4.51}$$

where the bar denotes an average over a period and

$$p(\eta) = \int_0^\eta \left[\frac{1}{c(s)} - \left(\overline{\frac{1}{c}}\right)\right] ds.$$

Hence

$$\phi = f\left(\left(\overline{\frac{1}{c}}\right)x + \varepsilon p\left(\frac{x}{\varepsilon}\right) - t\right), \tag{4.52}$$

and, neglecting terms of $O(\varepsilon)$, we see that, to the first order, the wave propagates as if it were in a *homogeneous* medium which has wave speed $(\overline{1/c})^{-1}$.

If we had not been able to solve (4.50) exactly, we could have used the method of multiple scales as described in [32] and [26] to get the same result. This method begins by regarding ϕ not just as a function of x and t, but as a function of x, X and t, where $X = x/\varepsilon$ is the second "scale". Thus, we generalise (4.50) in the form[13]

$$\frac{\partial \phi}{\partial t} + c(X) \left(\frac{\partial \phi}{\partial x} + \frac{1}{\varepsilon} \frac{\partial \phi}{\partial X} \right) = 0$$

for $\phi(x, X, t)$, where we will only be interested in the solution restricted to the plane $x = \varepsilon X$. We now seek a solution of the form

$$\phi \sim \phi_0 + \varepsilon \phi_1 + \cdots$$

and equate coefficients of ε to find

$$c(X) \frac{\partial \phi_0}{\partial X} = 0,$$

so that ϕ_0 is just a function of x and t. The next order terms give

$$c(X) \frac{\partial \phi_1}{\partial X} = -\frac{\partial \phi_0}{\partial t} - c(X) \frac{\partial \phi_0}{\partial x},$$

which can be integrated to give

$$\phi_1 = -X \frac{\partial \phi_0}{\partial x} - \frac{\partial \phi_0}{\partial t} \int_0^X \frac{dX'}{c(X')} + \phi_{10}(x, t) \tag{4.53}$$

for some function ϕ_{10}. Now we see that ϕ_1 is in danger of growing linearly with X which would invalidate the implicit assumption that $|\varepsilon \phi_1| \ll |\phi_0|$. To avoid this, ϕ_0 must satisfy the *secularity condition*

$$\frac{\partial \phi_0}{\partial t} + \frac{1}{\overline{(1/c)}} \frac{\partial \phi_0}{\partial x} = 0, \tag{4.54}$$

which is in accordance with result (4.52). Hence, (4.53) now implies that ϕ_1 differs from a function of (x, t) by $-p(X) \partial \phi_0 / \partial t$, confirming that waves on the X-scale are only apparent at the first order in ε.

We note that the characteristics of (4.54) are straight lines with slope $dx/dt = [\overline{(1/c)}]^{-1}$ which approximates the oscillatory characteristics of the exact equation.

The method of multiple scales can now be applied to the wave equation

$$(c(X))^2 \frac{\partial^2 \phi}{\partial x^2} = \frac{\partial^2 \phi}{\partial t^2}$$

---

[13] Note that we are implicitly assuming that the initial data is dependent only on x and not on X. If the initial data was just a function of X, ε could be removed from the problem by rescaling x and t, and the method of multiple scales could then tell us about the far field as $x, t \to \infty$.

in exactly the same way and we find that $\phi_0 = \phi_0(x, t)$, $\phi_1 = \phi_1(x, t)$ and

$$\frac{\partial^2 \phi_2}{\partial X^2} = \frac{1}{(c(X))^2} \frac{\partial^2 \phi_0}{\partial t^2} - \frac{\partial^2 \phi_0}{\partial x^2}$$

and hence the secularity condition implies that

$$\frac{\partial^2 \phi_0}{\partial x^2} = \left(\overline{\frac{1}{c^2}}\right) \frac{\partial^2 \phi_0}{\partial t^2}.$$

When using this method, we repeat that it is important to remember that the initial and boundary data must be independent of ε and that the layering must be periodic. Also it is clear that the homogenised equation is not valid everywhere because it predicts that signals travel at a speed $[(\overline{1/c^2})]^{-1/2}$ which is less than the average characteristic speed $[(\overline{1/c})]^{-1}$. It can also be shown that, for times and distances of $O(\varepsilon^{-2})$, the homogenised equation becomes dispersive [50].

Finally, we mention that wave propagation in randomly layered media, which is an important topic in the oil industry, for example, can lead to homogenised models in which there is damping and consequent localisation of the waves; this is known as "Anderson localisation" [27]. In this context, we recall that it was the effect of randomness that led to Landau damping in Section 4.4.2.

4.7 Stationary Waves

There are two frequency domain situations that are much easier to analyse than the general cases considered in the previous section. These occur when the wavelength $c/\omega = k^{-1}$ is either very large or very small compared to a typical length scale of interest, L, and in either case we can use asymptotic methods to analyse a number of problems. The first case occurs, for example, in acoustics, where wavelengths may be in metres and the second case arises frequently in electromagnetism, especially in optics, where the wavelengths may be about 10^{-9} metres. The case $\omega = 0$ corresponds to stationary waves and we start by considering k small enough that Helmholtz' equation (4.45) is a *regular perturbation*[14] of Laplace's equation. However, this limit does not usually reveal any very interesting phenomena unless there is a large source of energy that can be tapped, as is the case when the wave-bearing medium is moving bodily, and all the following problems fall into this category.

Hence, we will now revisit some examples that were introduced in Chapter 3 and consider surface gravity waves on a uniform stream and acoustic waves in a pipe of slowly varying cross-section and in a medium flowing past a thin obstacle. The consideration of problems in which k is large will be left to Section 4.8.

---

[14] For a definition of a regular perturbation, see Hinch [26].

4.7.1 Stationary Surface Waves on a Running Stream

We have seen in Exercise 3.4 that the dispersion relation for two-dimensional gravity waves on the surface of a stream of depth h moving with constant speed U is

$$(\omega \pm Uk)^2 = gk \tanh kh,$$

and it was also shown that stationary waves with $\omega = 0$ can occur only if $U^2 < gh$, in which case the flow is said to be *subcritical*. Waves of this type can often be observed upstream of an obstruction in a river and such waves will only be independent of time in a fairly sluggish flow.

The parameter $U/\sqrt{gh}$ is important in many free surface flows and it is called the *Froude Number*, F. Flows with $F > 1$ are called *supercritical*.

The solution is even more interesting in three dimensions when stationary waves with wave number k_1 in the **U** direction and wave number k_2 in a perpendicular direction satisfy the relation

$$U^2 k_1^2 = g|\mathbf{k}| \tanh |\mathbf{k}| h, \tag{4.55}$$

where $|\mathbf{k}| = (k_1^2 + k_2^2)^{1/2}$ (Exercise 3.5). Now, even in the simplest case of infinitely deep water, we can find many real values for **k** given U and g. When $h \to 0$, the equation reduces to

$$(U^2 - gh)k_1^2 = ghk_2^2,$$

and, as long as $k_2 \neq 0$, there will now be real solutions for **k** only in supercritical flow, when $U^2 > gh$. We will encounter this dispersion relation again in Section 4.7.3 when we study stationary waves in supersonic gas dynamics.

The beautiful pattern of waves behind a ship can be analysed using the dispersion relation (4.55). If a ship is travelling with speed U on deep water, then the general solution for the wave elevation is

$$\eta = \int_{-\infty}^{\infty} F(k_1) e^{-i(k_1 x + k_2 y)} \, dk_1, \tag{4.56}$$

where x and y are measured, from the ship, along and perpendicular to the direction of travel and, from (4.55) with $h \to \infty$, k_2 is related to k_1 by

$$U^2 k_1^2 = g(k_1^2 + k_2^2)^{1/2}; \tag{4.57}$$

hence, the integral in (4.56) is taken over values of k_1 satisfying[15] $|k_1| > g/U^2$. The function F will depend on the flow in the vicinity of the ship, but its precise form is unimportant when we look at waves far from the ship. This is because we can again use the method of stationary phase (Section 4.3) to estimate the form of η for large values of x and y. On writing $y = \lambda x$, and letting $x \to \infty$ while keeping λ constant,

---

[15] The contribution to the integral corresponding to complex k_2 can be shown to be negligible as $x, y \to \infty$.

we can apply (4.20) to see that the dominant contribution to the integral in (4.56) will come from values of k_1 for which

$$\frac{d}{dk_1}(k_1 + \lambda k_2) = 0. \tag{4.58}$$

From (4.57),

$$k_2 = \pm k_1 \left(\frac{U^4 k_1^2}{g^2} - 1\right)^{1/2},$$

and so (4.58) leads to

$$\lambda = \mp \left(\frac{U^4 k_1^2}{g^2} - 1\right)^{1/2} \Big/ \left(\frac{2U^4 k_1^2}{g^2} - 1\right). \tag{4.59}$$

This gives real values for k_1 only if $|\lambda| < 1/2\sqrt{2}$ and so we can see immediately that the waves are all contained in a wedge behind the ship of angle $\tan^{-1}(1/2\sqrt{2}) = \sin^{-1}(1/3)$. The angle of this wedge is thus independent of the speed or any other properties of the ship. The pattern of the wave crests, sketched in Figure 4.6, can also be calculated. All we need to do is to plot the level curves of the phase $k_1 x + k_2 y$ remembering that k_1 and k_2 are related via (4.57) and that k_1 is a function of $\lambda = y/x$ from (4.59). The details are left to Exercise 4.16. The edge of the wedge shown in Figure 4.6 is a kind of "envelope" of these wave crests or wavefronts, and we will see another example of this phenomenon in Section 4.8.

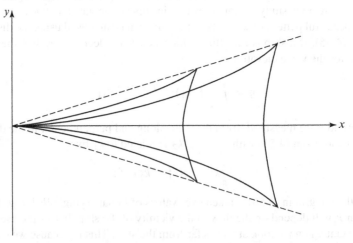

Fig. 4.6 Wavecrest pattern for ship waves

4.7.2 Steady Flow in Slender Nozzles

We next consider the homentropic, irrotational, steady flow of a gas along a pipe or nozzle which is aligned with the x-axis and is such that the cross-sectional area of the pipe $A(x)$ varies slowly with x. We could proceed by studying the steady form of (2.6), (2.7) and (2.9) and linearising about a unidirectional flow but it is easier, if less systematic, to revert to a "control volume"[16] approach (Figure 4.7). Assuming that the flow is unidirectional to a first approximation, we use conservation of mass on the volume shown in Figure 4.7 to say that

$$\rho u A = \text{constant.} \tag{4.60}$$

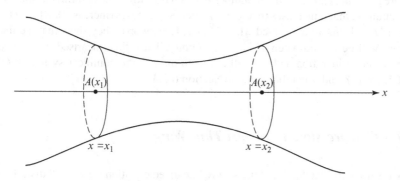

Fig. 4.7 Control volume for flow in a nozzle.

Also, by Bernoulli's equation (2.30),

$$\frac{1}{2}u^2 + \gamma p/(\gamma - 1)\rho = \text{constant,} \tag{4.61}$$

where we have used the result

$$\frac{p}{p_0} = \left(\frac{\rho}{\rho_0}\right)^{\gamma}, \tag{4.62}$$

since the flow is homentropic. Moreover, since the flow is irrotational, u is approximately a function of x and, hence, so are ρ and p. If we now linearise this system by writing $A = A_0 + \bar{A}$, $\rho = \rho_0 + \bar{\rho}$, $u = u_0 + \bar{u}$, $p = p_0 + \bar{p}$, where A_0, ρ_0, u_0, p_0 are all constants and $\bar{A}, \bar{\rho}, \bar{u}, \bar{p}$ are small perturbations, we get

$$\frac{\bar{\rho}}{\rho_0} + \frac{\bar{A}}{A_0} + \frac{\bar{u}}{u_0} = 0$$

---

[16] This is a very useful engineering technique in which the fluid is divided into regions, possibly large ones, and estimates are made for the *global* changes in mass, momentum and energy in these regions in terms of their boundary values.

and

$$u_0 \bar{u} + \frac{\gamma}{\gamma - 1} \frac{\bar{p}}{\rho_0} - \frac{c_0^2 \bar{\rho}}{(\gamma - 1)\rho_0} = 0.$$

Also, from (4.62), $\bar{p} = c_0^2 \bar{\rho}$, where $c_0^2 = \gamma p_0/\rho_0$, and so, solving for $\bar{\rho}, \bar{u}$ gives

$$\frac{\bar{\rho}}{\rho_0} = \frac{M^2 \bar{A}}{A_0(1 - M^2)}$$

and

$$\frac{\bar{u}}{u_0} = \frac{\bar{A}}{(M^2 - 1)A_0},$$

where $M = u_0/c_0$ is the Mach number of the basic flow. These formulae show that our intuition that u increases (decreases) when A decreases (increases), which is true when $M < 1$, must be reversed when $M > 1$. Even worse, they show that the linear theory will be invalid when M is close to unity. Thus, this linearised theory needs careful reconsideration and we will return to discuss the nonlinear system (4.60)–(4.62) for ρ, p and u in more detail in Section 6.2.3.

4.7.3 Compressible Flow past Thin Wings

It was shown in Chapter 3 that the steady linearised equation for a small disturbance to a uniform flow $U\mathbf{i}$ is (3.7), so that

$$M^2 \frac{\partial^2 \phi}{\partial x^2} = \nabla^2 \phi, \tag{4.63}$$

where $M = U/c_0$ is the Mach number of the undisturbed flow and $\varepsilon\phi$ is the velocity potential of the disturbed flow. The pressure is connected to ϕ via the relation

$$p = p_0 - \varepsilon \rho_0 U \frac{\partial \phi}{\partial x}, \tag{4.64}$$

where the small parameter ε characterises the size of the disturbance to the uniform flow. These equations can now be applied to the flow past a thin two-dimensional wing which is nearly aligned with the flow. We suppose the upper and lower surfaces of the wing are given by

$$y = \varepsilon f_\pm(x) \quad \text{for } 0 < x < l,$$

so that the boundary condition on the wing is

$$\varepsilon f'_\pm(x) = \frac{\varepsilon(\partial \phi/\partial y)}{U + \varepsilon(\partial \phi/\partial x)} \quad \text{on } y = \varepsilon f_\pm(x),$$

and the linear approximation reduces this to

$$\frac{\partial \phi}{\partial y} = U f'_\pm(x) \quad \text{on } y = 0 \pm \quad \text{for } 0 < x < l. \tag{4.65}$$

In order for the linearised model to be valid, we need to assume $f'_\pm(x)$ is $O(1)$.

As can be seen immediately from (4.63), the solution to this problem depends crucially on whether M is greater or less than 1.

(i) Subsonic Flow, $M < 1$

When M is less than 1, (4.65) is elliptic and, writing $\beta^2 = 1 - M^2$, we have to solve

$$\beta^2 \frac{\partial^2 \phi}{\partial x^2} + \frac{\partial^2 \phi}{\partial y^2} = 0 \tag{4.66}$$

subject to the boundary conditions (4.65) and the condition that $|\nabla \phi| \to 0$ at infinity. By rescaling $y = Y/\beta$, equation (4.66), becomes Laplace's equation and, if we also write $\phi = (1/\beta)\Phi(x, Y)$, then (4.65) remains

$$\frac{\partial \Phi}{\partial Y} = U f'_\pm(x) \quad \text{on } Y = 0 \pm.$$

Thus we see that Φ is the potential for an *incompressible* flow past the same thin wing.

It is easy to see that, since $\log(x^2 + \beta^2 y^2)$ is a solution of (4.66), a possible form for the solution of this problem is

$$\phi = \int_0^l g(\xi) \log((x - \xi)^2 + \beta^2 y^2) \, d\xi, \tag{4.67}$$

for some function g, which is equivalent to a source distribution along the x-axis.[17] Moreover, when we let $y \to 0+$ for $0 < x < l$, we find (Exercise 4.18) that

$$\frac{\partial \phi}{\partial y} \to 2\beta \pi g(x),$$

and so the choice

$$g(x) = \frac{U}{2\pi\beta} f'_+(x) \tag{4.68}$$

will make (4.67) satisfy the condition on the top of the wing.

It is only if the wing is symmetric, however, that the solution of the form (4.67) with g given by (4.68) will also satisfy the boundary condition as $y \to 0-$ and thus provide the solution we seek. For an asymmetric wing, we need to introduce a distribution of vortices as well as sources along the x-axis (Exercise 4.18).

---

[17] The logarithm in (4.67) is a Green's function for this problem [43].

This idea of using distributions of sources and vortices to represent a thin two-dimensional wing is exactly the same as that used in incompressible theory and, indeed, we can apply all the theory of incompressible flow past thin bodies to this problem. In particular, D'Alembert's paradox implies that there will be no forces on the wing if there is no circulation. Hence, for a symmetric wing there will be neither drag nor lift on the wing in subsonic flow. However, for an asymmetric wing with the Kutta–Joukowski condition applied at the trailing edge, there will be a lift, and we can relate the force in the compressible case to that in the incompressible case. In the linearised approximation, the lift L_c in the compressible case will be

$$L_c = \int_0^l (-p_+ + p_-)\, dx,$$

where $p_\pm$ are the pressures just above and below the wing. Using (4.64), we have

$$L_c = \varepsilon \rho_0 U \int_0^l \left(\frac{\partial \phi_+}{\partial x} - \frac{\partial \phi_-}{\partial x} \right) dx = \frac{1}{\beta} L_i, \tag{4.69}$$

where L_i is the lift in the incompressible case. For incompressible flow past a flat plate of length l at a small angle $-\varepsilon$ to the flow it can be shown (Acheson [1]) that the circulation is $\pi U l \varepsilon$ and hence, by the Kutta–Joukowski theorem, the lift on the wing in compressible flow will be $\rho_0 \pi U^2 l \varepsilon / \beta$, and there will be no drag.[18] This result is yet another manifestation of the breakdown of linear theory when $M \to 1$ and $\beta \to 0$.

(ii) Supersonic Flow, $M > 1$

When M is greater than 1, (4.63) is hyperbolic, and we can write down the general solution of this wave equation as

$$\phi = F(x - By) + G(x + By),$$

where $B^2 = M^2 - 1$. However, because the equation is hyperbolic, we need to impose different boundary conditions as compared to the subsonic case, and in particular we abandon the elliptic condition that $|\nabla \phi| \to 0$ at infinity. On physical grounds, we assert that there will be no *upstream influence* due to the wing, and therefore we impose Cauchy data $\partial \phi / \partial x = \phi = 0$ on $x = 0$. From this it follows that

$$\phi = \phi_+ = F(x - By) \ \text{ in } \ y > 0$$

and

$$\phi = \phi_- = G(x + By) \ \text{ in } \ y < 0.$$

---

[18] More generally, the lift on an arbitrary thin wing is $\rho_0 U \Gamma / \beta$, where Γ is the circulation around the wing in the incompressible case.

Now applying the boundary condition, (4.65) gives

$$-BF'(x) = Uf'_+(x), \quad BG'(x) = Uf'_-(x)$$

for $0 < x < l$, and the solution is

$$\phi_+ = -\frac{U}{B}f_+(x - By) \quad \text{for } 0 < x - By < l, \ y > 0 \tag{4.70}$$

and

$$\phi_- = \frac{U}{B}f_-(x + By) \quad \text{for } 0 < x + By < l, \ y < 0. \tag{4.71}$$

These formulas suggest that there will be "zones of silence" both in front of and behind the wing, as shown in Figure 4.8, and we will return to this point shortly.

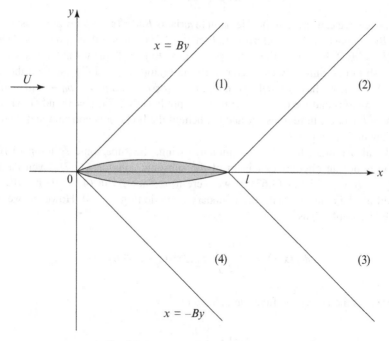

Fig. 4.8 Supersonic flow past a thin wing.

First, we use (4.69) to find the lift explicitly as

$$L = \frac{\rho_0 U \varepsilon}{B} \int_0^l (-f'_+(x) - f'_-(x)) \, dx$$

$$= -\frac{\rho_0 U^2 \varepsilon}{B}(f_+(l) + f_-(l))$$

assuming that $f_+(0) = f_-(0) = 0$. Thus, for a flat plate at incidence where $f_+(x) = f_-(x) = -x$, the lift will be non-zero and its value is $2\rho_0 U^2 \varepsilon l/B$, even without the assumption of a "trailing edge" condition. This is in marked contrast to the subsonic case, as is the fact that the drag on a flat plate now comes out to be $2\rho_0 U^2 \varepsilon^2 l/B$ (Exercise 4.21). The fact that the forces on the wing increase dramatically as $B \to 0$ or $M \to 1$ caused one of the major difficulties in the early days of supersonic flight.

It is well known that in incompressible flow, thin lifting wings leave wakes of concentrated vorticity behind them either in two-dimensional unsteady flow or in three-dimensional flow. To check that a supersonic wing does not shed such a wake even in steady flow, we return to the solution of (4.63) with $B^2 = M^2 - 1$. As remarked after (4.4) the equation can be written as

$$\left(B\frac{\partial}{\partial x} - \frac{\partial}{\partial y}\right)\left(B\frac{\partial}{\partial x} + \frac{\partial}{\partial y}\right)\phi = 0,$$

and hence, we can say that the Riemann invariants $B\partial\phi/\partial x \pm \partial\phi/\partial y$ are constant on $x \pm By = $ constant. Now, referring to Figure 4.8, we can see that in regions (1) and (2), $B\partial\phi/\partial x + \partial\phi/\partial y = 0$, so that $\phi = F(x - By)$ in (1) (as we already asserted) and also in (2). Similarly $\phi = G(x + By)$ in regions (4) and (3). Now on the line $y = 0, x > l$, we have to make sure that $v = \varepsilon\partial\phi/\partial y$ and $p = -p_0 - \varepsilon\rho_0 U\partial\phi/\partial x$ are continuous and we quickly see that this predicts that $F'(x) = 0$ and $G'(x) = 0$ for $x > l$. Hence, in regions (2) and (3) behind the body, ϕ is constant and there is no flow in these regions.

We also remark that, for a symmetric wing, the supersonic solution (4.70)–(4.71) is not an obvious extension of the subsonic solution (4.67). Even though $\log(x^2 - B^2y^2)$ satisfies (4.63), if we were to replace β^2 in (4.67) with $-B^2$, we would not be able to satisfy the boundary condition on $y = 0$. However, we can write (4.70)–(4.71) as

$$\phi_\pm(x, y) = \mp\frac{U}{B}\int_0^l f'_\pm(\xi)H(x - \xi \mp By)\,d\xi, \tag{4.72}$$

where H is the Heaviside function defined by[19]

$$H(x) = \begin{cases} 0, & x < 0 \\ 1, & x > 0. \end{cases}$$

Note that the upper limit of the integral in (4.72) is $x \mp By$ if $0 < x \mp By < l$, and there is no velocity perturbation downstream of the characteristics through the trailing edge.

This question of downstream influence is much more interesting, even for non-lifting bodies, in three dimensions, and we will consider the linearised flow past a

---

[19] This function is a Riemann function for this problem [43].

slender axisymmetric body such as a rocket in the next section. Before doing so, however, we must make two caveats about our linearised two-dimensional model. The first is that our predictions of infinite forces on aerofoils as $M \to 1$ in both subsonic and supersonic flow means that the linear model is invalid when $M^2 - 1$ is small. This fact is not surprising because, in this limit, the term $(M^2 - 1)\partial^2\phi/\partial x^2$ in (4.63) becomes so small as to be comparable with the nonlinear terms that we have neglected. Quite a complicated asymptotic procedure is needed to derive a consistent limit in this case and, because the upshot is a nonlinear model, we will defer its derivation to Section 6.3.1.

A similar caution applies when M is large, even though the slope of the body is of $O(\varepsilon)$, where ε is small. When $M\varepsilon$ is of $O(1)$, the region of influence of the body, which is bounded by $x = \pm By$ in two dimensions, becomes very thin. However, this inevitably introduces nonlinearity into the problem again, and so we leave further discussion to Section 6.3.3.

4.7.4 Compressible Flow past Slender Bodies

In this section, we consider the flow of a uniform stream past a slender axisymmetric body of length l given by $r = \varepsilon R(x)$ in cylindrical polars (r, θ, x). We will begin by considering a free stream which is aligned with the axis of the body and leave the case of a body at a small angle of incidence to the free stream to Exercise 4.20.

The equation to be solved is still (4.63), which can be written in cylindrical polar coordinates (r, θ, x) as

$$(M^2 - 1)\frac{\partial^2\phi}{\partial x^2} = \frac{\partial^2\phi}{\partial r^2} + \frac{1}{r}\frac{\partial\phi}{\partial r} + \frac{1}{r^2}\frac{\partial^2\phi}{\partial\theta^2}, \tag{4.73}$$

and we seek axisymmetric solutions in which $\phi = \phi(x, r)$. Since the velocity of the flow is $(0, 0, U) + \varepsilon\nabla\phi$, and the normal to the body is $(1, 0, -\varepsilon R'(x))$, the linearised form of the boundary condition on the body is

$$\frac{\partial\phi}{\partial r} = UR'(x). \tag{4.74}$$

In two dimensions, we obtained good results by applying this condition on the x-axis to give (4.65) and this certainly made the mathematics simpler. Now we will not be so lucky.

The need for modification becomes apparent when we regard the solution (4.67) as a distribution along $y = 0$, $0 < x < l$, of simple source solutions of equation (4.66) of the form $\log(x^2 + \beta^2 y^2)$. As shown in Exercise 4.18, this distribution leads to the integral in (4.67) having a Taylor series expansion in y for both $y > 0$ and $y < 0$, and this is what allows the boundary condition (4.65) to be applied on $y = 0$. When we try to distribute *axisymmetric* source solutions of (4.73), of the form $(x^2 + \beta^2 r^2)^{-1/2}$, along the line $r = 0$, $0 < x < l$, we find that $\partial\phi/\partial r$ inevitably approaches infinity as $r \to 0$. This is because the integral of $\partial\phi/\partial r$ around any small circle enclosing the body has to stay finite as the radius of the circle tends to zero. Thus, we will have to be careful when we apply the boundary condition (4.74).

(i) Subsonic Flow, $M < 1$

Having already observed that if $\beta^2 = 1 - M^2$, there is an elementary axisymmetric solution to equation (4.73) in the form $(x^2 + \beta^2 r^2)^{-1/2}$, we try a source distribution of the form

$$\phi = \int_0^l \frac{h(\xi)\, d\xi}{((x - \xi)^2 + \beta^2 r^2)^{1/2}}. \tag{4.75}$$

Now it can be shown (Exercise 4.18) that, as $r \to 0$,

$$\frac{\partial \phi}{\partial r} \sim -\frac{2h(x)}{r},$$

and so applying the boundary condition (4.74) on $r = \varepsilon R$, we find that

$$h(x) = -\frac{\varepsilon}{2} U R(x) R'(x). \tag{4.76}$$

Thus, we have obtained the solution to axisymmetric slender body theory in which the disturbance to the flow variables turns out to be of $O(\varepsilon^2)$ compared to the undisturbed quantities in spite of the fact that the body width is of $O(\varepsilon)$ compared to its length.

(ii) Supersonic Flow, $M > 1$

We recall that in going from subsonic flow to supersonic flow in two dimensions, we could not simply replace β^2 in (4.66) by $-B^2$. However, the fact that the boundary conditions in slender body flow have to be applied on $r = \varepsilon R$ rather than on $r = 0$ means that we may now be able to generalise (4.75) in order to solve the supersonic problem. Since the function

$$\psi = \begin{cases} 0 & |x| < Br \\ (x^2 - B^2 r^2)^{-1/2} & |x| > Br \end{cases} \tag{4.77}$$

formally satisfies (4.73), except when $|x| = Br$, we try

$$\phi = \int_0^{x - Br} \frac{m(\xi)\, d\xi}{[(x - \xi)^2 - B^2 r^2]^{1/2}} \tag{4.78}$$

for $x > Br$. The choice of the upper limit is not only suggested by the form of solution (4.77) but is also confirmed by Exercise 4.29 and by observing that the supersonic solution at $P(r, x)$ will only depend on the body shape upstream of point A, where the length OA is $x - Br$, as shown in Figure 4.9.

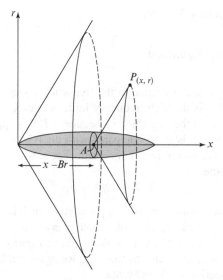

Fig. 4.9 Supersonic flow past a slender axisymmetric body.

Now, letting $r \to 0$ in (4.78), we find that $\phi \sim m(x) \cosh^{-1}(x/Br)$, which means that $\phi \sim m(x) \log r$ to lowest order; thus, using (4.74),

$$m(x) = \varepsilon U R(x) R'(x).\qquad(4.79)$$

Hence, the solution is

$$\phi = \int_0^{x-Br} \frac{\varepsilon U R(\xi) R'(\xi)\, d\xi}{((x-\xi)^2 - B^2 r^2)^{1/2}}$$

when $0 < x - Br < l$, and the limits of the integral will be 0 and l if $x - Br > l$. Thus, we see that although the body has no influence upstream of the characteristic cone $x = Br$ through the nose of the body, it does affect the flow *everywhere* inside this cone, even in $x - Br > l$. This is in contrast to the result for two-dimensional supersonic flow past a thin wing where the flow is influenced by the wing only *between* the characteristics or Mach lines through the leading and trailing edges of the wing. This is an example of *Huygens' Principle*, which will be discussed in more detail in Section 4.9. Also, we will say more about the implications of Figure 4.9 for the problem of sonic boom.

4.8 *High-frequency Waves

4.8.1 The Eikonal Equation

There is one other parameter regime in the frequency domain that is mathematically tractable and that is the high-frequency limit as ω or $k \to \infty$ in Helmholtz' equation. As explained at the beginning of Section 4.7, this approximation is relevant when the wavelength is small compared with the typical dimensions L of the region of interest.[20] We therefore nondimensionalise $\mathbf{x}$ with L so that Helmholtz' equation (4.45) becomes

$$\nabla^2 \Phi + k^2 \Phi = 0, \qquad (4.80)$$

where $k = \omega L/c$ is now a nondimensional parameter which is large for light waves in many everyday situations. We can look at both interior and exterior problems, but in any case, we expect that when $k \gg 1$, there will be *rapid spatial oscillations* in some or all of the region. Motivated by this idea, we again employ the WKB ansatz, which was used in Section 4.4.3, and write

$$\Phi \sim A(x, y)e^{iku(x,y)}, \qquad (4.81)$$

where we restrict ourselves to two dimensions for simplicity. As usual, u is the phase of Φ and A is the amplitude, and we remember that these variables are only uniquely defined in the limit as $k \to \infty$. In order to save ourselves a great deal of trouble, we will mostly confine ourselves to situations in which u is real. Substituting (4.81) into (4.80), we find that

$$\nabla \Phi \sim (ikA\nabla u + \nabla A)e^{iku},$$

and

$$\nabla^2 \Phi \sim (-k^2 A(\nabla u)^2 + 2ik\nabla A.\nabla u + ikA\nabla^2 u + \nabla^2 A)e^{iku},$$

so that as $k \to \infty$, the lowest order terms in (4.80) reveal the *eikonal equation*

$$|\nabla u|^2 = 1. \qquad (4.82)$$

The second-order terms will give the so-called transport equation

$$A\nabla^2 u + 2\nabla A.\nabla u = 0, \qquad (4.83)$$

which determines A. Hence, we have taken the possibly retrograde step of transforming a linear second-order equation (4.80) to a fully nonlinear first-order equation (4.82), whose consideration should perhaps be deferred to Chapter 5. Assuming the reader has some familiarity with nonlinear first-order scalar differential equations,

---

[20] Of course, if we are interested in waves sufficiently close to a sharp corner of some boundary, the method we are about to describe will never be useful, because then L can be arbitrarily small.

however, we will continue here because (4.82) can give us very helpful insights into high-frequency linear wave propagation.

We can use *Charpit's method* (see [43]) to see that the solution of (4.82) is given by solving the characteristic equations

$$\frac{dx}{d\tau} = 2p, \quad \frac{dy}{d\tau} = 2q, \quad \frac{du}{d\tau} = 2, \quad \frac{dp}{d\tau} = 0, \quad \frac{dq}{d\tau} = 0, \qquad (4.84)$$

where $p = \partial u/\partial x$ and $q = \partial u/\partial y$, and τ is a parameter that varies along the characteristics. Initial data is required for these equations and will be given by the boundary conditions imposed on Φ, which may be at infinity for an exterior problem. We note at once that since p and q are constant from (4.84), the characteristics will be straight lines.

The simplest solution of (4.82) is $u = lx + my$, where $l^2 + m^2 = 1$, and this solution represents a plane wave. Note that the relation $l^2 + m^2 = 1$ is the dispersion relation. More generally, it can be shown that if the high-frequency approximation to the dispersion relation for a problem is given by

$$\frac{\omega L}{c} = \Omega(L\mathbf{k}),$$

where $\mathbf{k}$ is the wave number as defined in Section 4.4, then the equation for u in the WKB approximation for the same problem will be

$$\Omega(\nabla(u)) = 1.$$

4.8.2 Ray Theory

Using the terminology of Section 4.5, the nicest interpretation of the characteristics, given by (4.84), of the eikonal equation (4.82) is as *rays*. For the plane wave $u = lx + my$, the rays are straight lines in the direction (l, m) which is perpendicular to the *wavefront* defined to be the curve on which u is constant. It can easily be seen from (4.84) that this geometric relation between the rays and the wavefronts always holds.

In the case of electromagnetism, these rays are the familiar light rays that are drawn in optics in elementary physics courses. Indeed, the well-known pictures for light reflected by a mirror (Figure 4.10(a)) or refracted by a lens (Figure 4.10(b)) are easily seen to be simple superpositions of solutions of the eikonal equation. These pictures result from the fact that the plane wave solution of the eikonal equation is an exact solution of Helmholtz' equation, and so it is straightforward to see that once we have imposed suitable boundary conditions on u, the rays of the eikonal equation reflect in a mirror exactly as predicted in Section 4.5.2. Similarly, for refraction in a lens, if the speed of sound in the refractive medium is c_1, then, with $k = \omega L/c_1$, the eikonal equation is

$$|\nabla u|^2 = 1$$

in the air and

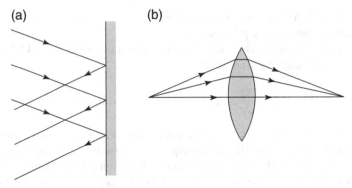

Fig. 4.10 (a) Light rays reflected from a plane mirror. (b) Light rays refracted by a lens

$$|\nabla u| = \left(\frac{c_0}{c_1}\right)^2$$

in the lens. When we impose continuity of u at the interface, this leads to straight line rays in each region joined together by using Snell's law of refraction (4.48).

A much more interesting solution of (4.82) arises if we consider waves inside a circle of radius $\sqrt{2}$, say, where perimeter is being oscillated so that

$$u = s \text{ on } x = \cos s + \sin s, \quad y = \sin s - \cos s \text{ for } 0 < s < 2\pi, \qquad (4.85)$$

and we assume k is an integer for ϕ to be 2π-periodic in s. It can be seen that on this circle, either $p = \cos s$, $q = \sin s$ or $p = -\sin s$, $q = \cos s$. Thus, solving equations (4.84) and applying these boundary conditions when $\tau = 0$ leads to two possible solutions:

(i) $u = 2\tau + s$ (ii) $u = 2\tau + s$

    $x = 2\tau \cos s + \cos s + \sin s$ $x = -2\tau \sin s + \cos s + \sin s$

    $y = 2\tau \sin s + \sin s - \cos s$ $y = 2\tau \cos s + \sin s - \cos s$

    $p = \cos s$ $p = -\sin s$

    $q = \sin s;$ $q = \cos s.$

These solutions are the two families of straight lines, shown in Figure 4.11, which intersect the circle in directions making angles $\pm\pi/4$ with the outward normal at each point. Thus, when $\tau = -\frac{1}{2}$, these rays will all touch their envelope, the circle $x^2 + y^2 = 1$, and this curve is called a *caustic*. We can verify that the transformation from (x, y) to (s, τ) is singular on the caustic since $J = \partial(x, y)/\partial(s, \tau)$ vanishes when $\tau = -\frac{1}{2}$.

The fact that caustics like this can be expected to occur in most problems can be seen without using any mathematics. If we consider planar light waves with $k \gg 1$ impinging on a curved surface, then the reflected rays will always be a family of straight lines which will, in general, have an envelope. If we consider, for example, the sun shining on the inside of a circular coffee cup, the caustic which can be seen

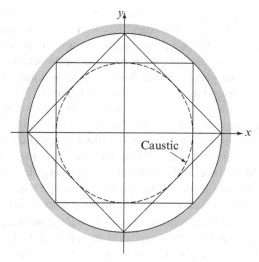

Fig. 4.11 Caustics and characteristics for the eikonal equational with data (4.85).

on the surface of the coffee will be a nephroid, and a partial ray picture is illustrated in Figure 4.12.

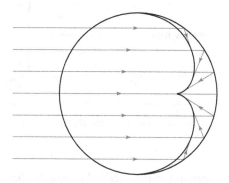

Fig. 4.12 The nephroid caustic on a cup of coffee.

In both the above solutions, families of rays apparently do not penetrate beyond the caustic. Thus, the eikonal equation predicts that there will be a light region separated from a dark region by the caustic, and an analysis of the equation (4.83) for A reveals that $|A| \to \infty$ as the caustic is approached, so that the maximum illumination is seen on the caustic itself. However, there cannot be any singularity in the solution Φ of Helmholtz' equation, which is an elliptic partial differential equation.

Before we resolve this difficulty, we mention another disturbing consequence of the fact that the eikonal equation can have real characteristics. This implies that almost any kind of singularity in the boundary data for Φ will, if k is large, cause a singularity in u to propagate away from the boundary even though Φ can have no singularities away from the boundary. However, this observation gives us the key to the success of the WKB approximation in so many situations in optics or acoustics. If we return to the diffraction problem mentioned in Section 4.5.2 for rays impinging on a flat plate, then the eikonal equation simply predicts a shadow with clear cut edges behind the plate, as shown in Figure 4.13(a). This is not a bad approximation to what happens in practice when $k \gg 1$. However, even with large k, Helmholtz' equation will predict diffraction at the edges A, B of the plate and, by careful analysis, it can be shown that diffracted fields propagate radially from A and B, as shown schematically in Figure 4.13(b). The strength of these fields is $O(k^{-1/2})$ relative to the incident wave field. On the other hand, the diffracted field generated by a smooth body is more subtle. Again, the high-frequency approximation predicts incident and reflected waves and a clear cut shadow region, as in Figure 4.14(a), but now it can be shown that the amplitude of the field in the shadow region is $O(e^{-k^{1/3}})$ compared to the incident field. This dark shadow indicated in Figure 4.14(b) needs to be described by rays which are known as *creeping rays* rather than the diffracted rays of Figure 4.13(b) (see Born and Wolf [6]).

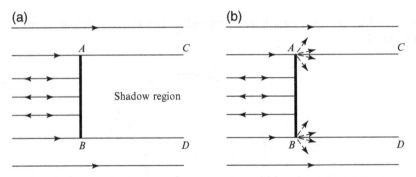

Fig. 4.13 (a) Shadow region for a flat plate using ray theory. (b) The effect of diffraction at the edges of a flat plate.

Both of the shadow regions can be described by real ray theory if we are prepared to undertake all the complicated "singular perturbation" analysis which is needed to unravel the structure of the solution near, say, the diffraction points A and B. More importantly, as described briefly in the next section, thin "boundary layers" can be constructed along AC, BD to smooth out the discontinuities between the incoming waves and the refracted field. This not only resolves the apparent contradiction between the hyperbolicity of the eikonal equation and the ellipticity of Helmholtz' equation, but it also gives us the clue as to what happens in the even deeper shadow within the central circle in Figure 4.11, where the wavefield appears to end in a caustic. In this case, a boundary layer analysis close to the caustic reveals that the field inside the caustic is exponentially small in k as $k \to \infty$ and that the rays in this region are *complex* and x, y, p, q and u all take complex values. Unfortunately, complex ray theory requires an even more elaborate asymptotic development than does real ray theory (see Chapman et al. [9]).

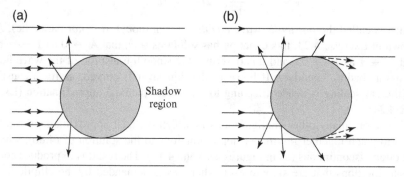

Fig. 4.14 (a) Rays reflected from a smooth body using ray theory. (b) The effect of diffraction for a smooth body.

We remark that if we solve the eikonal equation with the boundary condition (4.85) replaced by $u = \varepsilon s$ on the same circle, where $k\varepsilon$ is now an integer, we find that the caustic lies closer and closer to the boundary $x^2 + y^2 = 2$ as $\varepsilon \to 0$. Now, although the introduction of a boundary layer structure can smooth out the discontinuity on the caustic, most of the wave field will still be "trapped" close to the boundary, and this is an example of the famous *whispering gallery waves* that can easily be generated inside the dome of St Paul's Cathedral, for example. Another readily observed caustic, but one with quite a different structure, is the boundary $\sin \theta = \pm 1/2\sqrt{2}$ of the ship wave pattern described in Section 4.7.1.

4.8.3 Paraxial Approximations

It is interesting that many of the above-mentioned phenomena can be modelled more accurately than is possible by ray theory by seeking approximate solutions of Helmholtz' equation using the key assumption that, for large k, the wave field varies rapidly across a thin region; this could happen, say, for a torch beam or at the boundary of a shadow. The simplest configuration in which this idea can be used is when the thin layer is planar and perpendicular to the y-axis. To describe this *paraxial* situation in just two dimensions, we write

$$\Phi = Ae^{ikx}$$

in (4.80) to find

$$2ik\frac{\partial A}{\partial x} + \frac{\partial^2 A}{\partial x^2} + \frac{\partial^2 A}{\partial y^2} = 0.$$

Then using a boundary layer scaling $y = Y/\sqrt{k}$, and remembering that $k \gg 1$, the lowest order terms are

$$\frac{\partial^2 A}{\partial Y^2} + 2i\frac{\partial A}{\partial x} = 0, \tag{4.86}$$

which is called the *parabolic wave equation* (even though it is *not* parabolic). As shown in Exercise 4.23, this equation has solutions such that $A \to 0$ as $Y \to -\infty$ and $A \to 1$ as $Y \to \infty$, and this describes the smooth transition of the wave field across a shadow boundary. If however the thin layer is curved, as for a caustic, a different scaling is needed, leading to a different paraxial approximation (Exercise 4.24).

There are many other configurations in which paraxial approximations can be used to patch together ray theory approximations to the solution of (4.80), which are often discontinuous (as in Figures 4.13 and 4.14). The result is to produce composite solutions that are smooth everywhere, as is demanded by the ellipticity of equation (4.80), see [46]. These solutions form part of the *Geometrical Theory of Diffraction*.

4.9 *Dimensionality and the Wave Equation

At several points in this chapter, we have remarked on the way in which the *qualitative* nature of linear wave propagation as described by the wave equation (3.6) depends on the number of space dimensions. The most striking piece of evidence for this has been the observation in Sections 4.7.3 and 4.7.4 that a thin wing moving supersonically in two dimensions leaves behind it no wake at all, whereas a slender axisymmetric projectile can always be detected after its passage.

A more familiar scenario concerns the propagation of a disturbance that is localised near a point in space and time. We know that for the one-dimensional case of, say, waves on a string, an initial disturbance localised near $x = 0$ will eventually emerge as two pulses each propagating without change or diminution near $x = \pm ct$, and there will be no disturbance anywhere else. Similarly, using the solution (4.6) in three dimensions, an initial disturbance near $r = 0$ will evolve into one that is localised in the space (x, y, z, t) near the spherical "shell" $x^2 + y^2 + z^2 = c^2 t^2$, albeit with a decay factor of r^{-1}. This is in accord with the evidence from our eardrums, but, when we look at a disturbance initially localised on the surface of a drum or created by dropping a pebble into a pond, the situation is quite different.

To appreciate this difference, we first remark that the localisation of the waves in the string example mentioned above is identical to the localisation of the waves emitted by the supersonic wing as given by (4.70) and (4.71) in the limit $l \to 0$. We simply have to identify x with time and y with distance along the string to make the problems mathematically equivalent. Now we are in a position to understand the evolution of two-dimensional axisymmetric waves satisfying

$$\frac{\partial^2 \phi}{\partial r^2} + \frac{1}{r}\frac{\partial \phi}{\partial r} = \frac{1}{c^2}\frac{\partial^2 \phi}{\partial t^2} \tag{4.87}$$

by drawing a similar analogy with the steady axisymmetric supersonic flow problem described in Section 4.7.4 when $M > 1$. When we again let the length of the body tend to zero, which is equivalent to the release of a short pulse of sound at $t = 0$ in (4.87), we find that the solution (4.77) gives us

$$\phi = \begin{cases} 0 & r > ct \\ \dfrac{\lambda}{(c^2t^2 - r^2)^{1/2}}, & r < ct, \end{cases} \tag{4.88}$$

for some constant λ. Hence, although there is a sharp front at $r = ct$ as in the one- and three-dimensional cases, in two dimensions the disturbance is felt everywhere inside the cone $r = ct$ and is *not* localised near the cone $r = ct$. Those who doubt this argument may ask themselves why a lightning strike, which may be approximated as an instantaneous energy release along a vertical line, produces *rumbles* of thunder after the sharp crack at $r = ct$.

The mathematics can be beautifully unified by the theory of the *retarded potential*. It can be shown [43] that the solution of the three-dimensional wave equation with an initial condition $\partial\phi/\partial t = f(x, y, z)$, $\phi = 0$ is given by the formula

$$\phi(x, y, z, t)$$
$$= \frac{ct}{4\pi} \int_0^{2\pi} \int_0^{\pi} f(x + ct\sin\theta\cos\phi, y + ct\sin\theta\sin\phi, z + ct\cos\theta)\sin\theta\, d\theta\, d\phi. \tag{4.89}$$

It can be verified (after much work) that this function satisfies the wave equation and, in fact, a linear combination of ϕ and $\partial\phi/\partial t$ can be used to satisfy any given initial conditions. Now the physical significance of (4.89) is that if f is zero outside a region D, then the integral will be non-zero only for times $t_{min} < t < t_{max}$ where, as shown in Figure 4.15, t_{min} is the first time at which the disturbance is felt at (x, y, z), and t_{max} is the last time.

We see at once that if the initial disturbance is very elongated in the z direction, then t_{max} will tend to infinity while t_{min} remains fixed. Hence, we have *Huygens' principle* that there are sharp "leading" and "trailing" wavefronts in one or three dimensions whereas in two dimensions there is only a sharp leading wave front.

We must emphasise that the above discussion only applies to the wave equation with constant coefficients. For waves in inhomogeneous media (including three-dimensional waves in a half space) or for waves governed by models other than (3.6), there is no reason for Huygens' principle to hold. What is true generally is that the waves decay with distance more rapidly as the number of space dimensions increases. However, it is no easy matter to estimate this rate of decay for a general wave model.

The above discussion can give us insight into the possible evolution of the steady wave patterns considered in Sections 4.7.3 and 4.7.4. The expression (4.89) is the general solution of

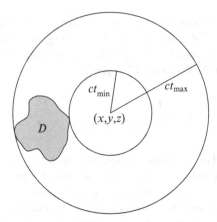

Fig. 4.15 The effect of a 3-dimensional disturbance initially localised in D.

$$\frac{\partial^2 \phi}{\partial x^2} + \frac{\partial^2 \phi}{\partial y^2} + \frac{\partial^2 \phi}{\partial z^2} = \frac{1}{c^2} \frac{\partial^2 \phi}{\partial t^2},$$

which is sometimes called the wave equation in the *acoustic frame*. However, we set $\xi = x - Ut$ and consider solutions of the wave equation in the *aerodynamic frame*, namely,

$$\left(1 - \frac{U^2}{c^2}\right) \frac{\partial^2 \phi}{\partial \xi^2} + \frac{\partial^2 \phi}{\partial y^2} + \frac{\partial^2 \phi}{\partial z^2} = \frac{1}{c^2} \frac{\partial^2 \phi}{\partial t^2} - \frac{2U}{c^2} \frac{\partial^2 \phi}{\partial \xi \partial t};$$

in Sections 4.7.3 and 4.7.4 we restricted ourselves to the "steady" case, where $\phi = \phi(\xi, y, z)$.

To see how such a steady flow might be set up, let us consider, in the acoustic frame, what happens when a two-dimensional disturbance is localised near $x = Ut$, $y = 0$ for $t \geq 0$ and with $U > c$. The solution at time t consists of the super-position of the solutions generated at times τ, where $0 < \tau < t$, and the above discussion reveals that the contribution from time τ is contained within the cone $(x - U\tau)^2 + y^2 = c^2(t - \tau)^2$ in (x, y, t) space. From (4.88), the amplitude of this contribution is non-zero inside this cone and is infinite at the cone surface which is sometimes called the *wavefront*[21] (yet another use of the term). Hence, the solution at time t in the acoustic frame is contained within the superposition of the cones, as shown in Figure 4.16a. The projection of the "tops" of the cones in the aerodynamic (x, y) plane is simply the sequence of circles shown in Figure 4.16b beginning from the "starting" circle $x^2 + y^2 = c^2t^2$. We thus see the following:

(i) The characteristics in the aerodynamic frame, $\xi = \pm y\sqrt{U^2/c^2 - 1}$ emerge as the *envelope* of the wavefronts in the acoustic frame.

---

[21] In spatial dimension $n \geq 2$, the characteristic surfaces or wavefronts of a hyperbolic system can be defined by generalising the ideas of Section 4.1. They are the only manifolds of dimension $(n - 1)$ across which jumps in the variables can occur and these jumps can be localised along bicharacteristic curves or "rays" in the characteristic manifold as described in [43].

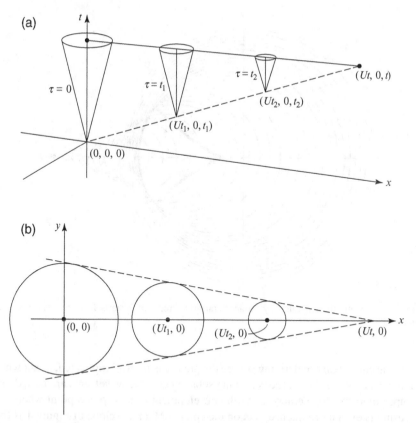

Fig. 4.16 (a) The influence of a supersonic point source of sound at $x = Ut$, $y = 0$ in (x, y, t) space, (b) The influence of a supersonic point source in (x, y) space.

(ii) There is a complicated non-zero disturbance between these characteristics for all finite time. However, as $t \to \infty$ this disturbance decays through destructive interference, leaving the "no-wake" flow described in Section 4.7.3.

We conclude this section with a brief discussion of the implications of the above scenario for the problem of *sonic boom* caused by supersonic aircraft. A sonic boom is a sudden jump in pressure which can usually only be described by the nonlinear theory in Chapter 6. However, many commonly occurring boom effects can be understood by modelling the phenomena as singular solutions of the linear acoustic equations as in Figures 4.9, 4.13, 4.14 and 4.16. In order to describe the transition from subsonic to supersonic flow, we consider the wave fronts emanating from a point source, as modelled by a delta function on the right-hand side of (4.87), moving along the x-axis as it accelerates through the speed of sound. Figure 4.16(b) depicts what happens when steady supersonic flow has been attained and the disturbance is contained within a cone or a wedge. However, when the source is moving subsonically, the wave fronts are all contained within the wavefront that emanated from the source at $t = 0$.

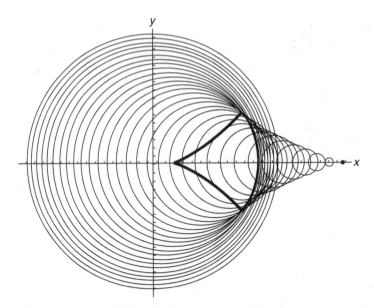

Fig. 4.17 Wavefronts generated by an accelerating source which starts from rest and becomes supersonic.

In the case of an accelerating source the picture is more complicated, as shown in Figure 4.17. While the source is moving subsonically, the wavefronts are nested, but a supersonic envelope emerges via the development of a cusp at a point where the focusing is even more intense than on the conical Mach envelope of Figure 4.16(b). From the three-dimensional version of this picture, it is possible to take a horizontal slice and thus to predict the "sonic carpet" felt at ground level due to a supersonic aeroplane (see [31] for more details).

Exercises

4.1 (i) It can be shown that, for steady small two-dimensional disturbances on a uniform flow in a weakly stratified gas with Mach number M, the velocity potential ϕ of the disturbance satisfies

$$(M(y))^2 \frac{\partial^2 \phi}{\partial x^2} = \frac{\partial^2 \phi}{\partial x^2} + \frac{\partial^2 \phi}{\partial y^2}.$$

If $M = (1+y)^{1/2}$, in which case the equation is called a *Tricomi equation*, show that the characteristics are

$$\pm 2y^{3/2} = 3x + \text{constant.}$$

Sketch these characteristics and indicate the "region of influence" of a small obstacle which is put in the flow at a point where $y > 0$.

(ii) Suppose that, in the same stratified flow, a wave of the form $\phi = \text{Rl}\,(e^{ikx}A(y))$ is incident on $y > 0$ from above. Show that

$$\frac{d^2 A}{dy^2} + k^2 yA = 0.$$

This is *Airy's equation* and it shows that ϕ will be oscillatory in $y > 0$, but that, in $y < 0$, it will decay exponentially. Thus, waves from above will not penetrate far into $y < 0$ and the disturbance due to the obstacle in part (i) will decay exponentially in $y < 0$.

R 4.2 In axisymmetric acoustic wave propagation inside a rigid circular cylinder of radius a, the velocity potential $\phi(r, t)$ satisfies

$$\frac{1}{c^2}\frac{\partial^2 \phi}{\partial t^2} = \frac{\partial^2 \phi}{\partial r^2} + \frac{1}{r}\frac{\partial \phi}{\partial r} \quad \text{for } 0 \le r < a,$$

with

$$\frac{\partial \phi}{\partial r} = 0 \quad \text{on } r = a.$$

Show that ϕ can be written as a generalised Fourier series in the form

$$\phi = \sum_{n=0}^{\infty}(a_n \cos \omega_n t + b_n \sin \omega_n t)J_0\left(\frac{\omega_n r}{c}\right),$$

where $J_0(z)$ satisfies

$$\frac{d^2 J_0}{dz^2} + \frac{1}{z}\frac{dJ_0}{dz} + J_0 = 0, \tag{$\dagger$}$$

with $J_0(0) = 1$, ω_n are defined by $J_0'(\omega_n a/c) = 0$, and a_n, b_n are arbitrary constants. It can be shown[22] that

$$\int_0^a r J_0\left(\frac{\omega_n r}{c}\right) J_0\left(\frac{\omega_m r}{c}\right) dr = \begin{cases} 0, & m \neq n \\ \frac{1}{2}a^2\left[J_0\left(\frac{\omega_n a}{c}\right)\right]^2, & m = n \end{cases}.$$

Show that if the gas in the tube is initially at rest with a pressure distribution $P_0(r)$, then

$$\phi(r,0) = 0 \quad \text{and} \quad \frac{\partial \phi}{\partial t}(r,0) = -\frac{1}{\rho_0}P_0(r).$$

Hence, show that $a_n = 0$ and that

$$b_n = \frac{-2}{\rho_0 a^2 \omega_n} \frac{\int_0^a r P_0(r) J_0(\omega_n r/c)\, dr}{[J_0(\omega_n a/c)]^2}.$$

4.3 Consider one-dimensional acoustic waves propagating with speed c in an organ pipe which is closed at $x = 0$ and open to the atmosphere at $x = L$. Assuming constant pressure at $x = L$, show that the velocity potential ϕ satisfies

$$\frac{\partial \phi}{\partial x}(0,t) = 0 \quad \text{and} \quad \phi(L,t) = 0.$$

If $\phi(x,0) = f(x)$ and $\partial\phi(x,0)/\partial t = g(x)$, show that

$$\phi = \sum_{n=0}^{\infty}\left(a_n \cos\frac{(2n+1)\pi ct}{2L} + b_n \sin\frac{(2n+1)\pi ct}{2L}\right)\cos\frac{(2n+1)\pi x}{2L},$$

where

$$a_n = \frac{2}{L}\int_0^L f(x) \cos\frac{(2n+1)\pi x}{2L}\, dx$$

and

$$b_n = \frac{4}{(2n+1)\pi c}\int_0^L g(x) \cos\frac{(2n+1)\pi x}{2L}\, dx.$$

This could be a model for a motor bicycle exhaust, so suppose, more realistically, that the open end radiates sound to the environment. One model for this is to say that, at $x = L$,

$$\frac{\partial \phi}{\partial t} + \alpha \frac{\partial \phi}{\partial x} = 0.$$

---

[22] To prove this orthogonality result, put $z = \omega_n r/c$ in (†) and then multiply by $2r^2 J_0'(\omega_n r/c)$ and integrate from $r = 0$ to a, integrating by parts once.

Show that if we assume that high pressure in the pipe pumps gas into the
environment and low pressure sucks gas into the pipe, then $\alpha > 0$.

By separating the variables in equation (4.1), show that

$$\phi = (A \cos \omega_n t + \beta \sin \omega_n t) \cos \frac{\omega_n x}{c},$$

where

$$\tan^2 \frac{\omega_n L}{c} + \frac{c^2}{\alpha^2} = 0.$$

Deduce that ω_n is complex and show that, for $\alpha \ll c$,

$$\omega_n = \frac{(2n+1)\pi c}{2L} \pm i\frac{\alpha}{L}.$$

Show that the solution of the initial value problem is now

$$\phi = \sum_{n=0}^{\infty} e^{-\alpha t/L}\left(a_n \cos \frac{(2n+1)\pi ct}{2L} + b_n \sin \frac{(2n+1)\pi ct}{2L}\right)\cos \frac{(2n+1)\pi x}{2L},$$

where a_n and b_n are as given above.

*4.4 Show that the Fourier transform of $\varepsilon/(x^2 + \varepsilon^2)$ is $\pi e^{-\varepsilon|k|}$.

If the impact of a long stick on a pond is modelled by taking the initial
surface profile to be $\eta_0(x) = -a\varepsilon/\pi(x^2 + \varepsilon^2)$, show that

$$\eta(x, t) = -\frac{a}{\pi} \int_0^{\infty} e^{-\varepsilon k} \cos \sqrt{gk} t \cos kx \, dk.$$

Use the method of stationary phase to show that, as $x, t \to \infty$, the major
contribution to the integral comes from values of k satisfying

$$x \pm \frac{1}{2}\sqrt{\frac{g}{k}}t = 0,$$

and hence that, for $x > 0$,

$$\eta \sim -\frac{at}{2}\sqrt{\frac{g}{\pi x^3}} \cos\left(\frac{gt^2}{4x} - \frac{\pi}{4}\right)$$

when ε is sufficiently small.

Which of the following pictures in Figure 4.18 is the more realistic representation of this long-time behaviour?

(a)   (b)

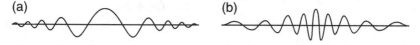

Fig. 4.18 Possible waves generated by a localised source.

4.5 As will be shown in Section 5.2.3, tidal waves can be modelled by the wave equation

$$gh\left(\frac{\partial^2 \eta}{\partial r^2} + \frac{1}{r}\frac{\partial \eta}{\partial r} + \frac{1}{r^2}\frac{\partial^2 \eta}{\partial \theta^2}\right) = \frac{\partial^2 \eta}{\partial t^2},$$

where η is the height of waves on the surface of an ocean of constant depth h and r, θ are polar coordinates in the horizontal plane.

A tsunami is simulated by supposing that at $t = 0$, $\eta = e^{-r^2/2}$ and $\partial \eta / \partial t = 0$. You are given that the inversion formulae for the *Hankel Transform* $\hat{f}(k)$ of the function $f(r)$ are

$$\hat{f}(k) = \int_0^\infty rJ_0(kr)f(r)\,dr$$

and

$$f(r) = \int_0^x kJ_0(kr)\hat{f}(k)\,dk,$$

where $J_0(x)$ is defined in Exercise 4.2. Show that the tidal wave equation transforms into

$$\frac{d^2\hat{\eta}}{dt^2} = -c^2 k^2 \hat{\eta},$$

where $gh = c^2$ (see [43] if you need help). Hence, show that

$$\eta(r,t) = \int_0^\infty \rho e^{-\rho^2/2}\left(\int_0^\infty kJ_0(kr)J_0(k\rho)\cos(kct)\,dk\right)d\rho,$$

which can be reduced to a single integral.

4.6 The effect of a constant surface tension T on two-dimensional interfacial gravity waves is to introduce a pressure drop of T/R across the interface $y = \eta(x,t)$, where R is the local radius of curvature of the interface. Show that, with a sign convention that you should specify, R^{-1} is approximately $\partial^2 \eta / \partial x^2$ for small amplitude waves.

Show that if a liquid of density ρ_1, pressure p_1 lies above a liquid of density ρ_2, pressure p_2, with the interface being given by $z = \eta(x,t)$, then

$$p_1 - p_2 = T\frac{\partial^2 \eta}{\partial x^2} \quad \text{on } z = 0,$$

to lowest order. Deduce that if $\eta = a\cos(kx - \omega t)$, the dispersion relation is

$$\omega^2 = g|k|\frac{(\rho_2 - \rho_1)}{(\rho_2 + \rho_1)} + \frac{T|k|^3}{(\rho_2 + \rho_1)}.$$

This implies that surface tension can stabilise such an interface even if $\rho_1 > \rho_2$. Is surface tension more effective as a stabilising mechanism for large $|k|$ or for small $|k|$?

*4.7 The displacement $\mathbf{u}$ in an elastic medium satisfies equation (3.72), where $\mathbf{X} = \mathbf{u}$. If $\mathbf{u}$ is written as $\mathbf{u} = \mathrm{Rl}\, \mathbf{u}_0 e^{i(\mathbf{k}.\mathbf{x} - \omega t)}$, where $\mathbf{u}_0 = A\mathbf{k} + \mathbf{B} \wedge \mathbf{k}$, with $A, \mathbf{k}$ and $\mathbf{B}$ all constant, show that either $\mathbf{B} = 0$ and $(\lambda + 2\mu)|\mathbf{k}|^2 = \rho\omega^2$, or $A = 0$ and $\mu|\mathbf{k}|^2 = \rho\omega^2$. If $k_p = \omega\sqrt{\rho/(\lambda + 2\mu)}$ and $k_s = \omega\sqrt{\rho/\mu}$, and c_p and c_s are the corresponding wave speeds, it can be shown that $c_p > c_s$.

The boundary conditions for a two-dimensional displacement $\mathbf{u} = (u(x,y,t), v(x,y,t), 0)$ in $y < 0$, with an unstressed boundary at $y = 0$, are

$$\frac{\partial u}{\partial y} + \frac{\partial v}{\partial x} = 0 = 2\mu\frac{\partial v}{\partial y} + \lambda\left(\frac{\partial u}{\partial x} + \frac{\partial v}{\partial y}\right),$$

(see (3.73) and Exercise 3.14). Show that a solution for a propagating wave in the form

$$\mathbf{u} = \mathrm{Rl}\,(\mathbf{a}_p e^{\kappa_p y} + \mathbf{a}_s e^{\kappa_s y}) e^{i(kx - \omega t)},$$

where $\kappa_p^2 = k^2 - k_p^2$ and $\kappa_s^2 = k^2 - k_s^2$, is possible as long as

$$(2 - \frac{c^2}{c_s^2})^2 = 4(1 - \frac{c^2}{c_p^2})^{1/2}(1 - \frac{c^2}{c_s^2})^{1/2},$$

where $c = \omega/k < c_s$. This wave is called a *Rayleigh wave*.

*4.8 From Exercise 3.1, the perturbation potential ϕ for small two-dimensional disturbances in a uniform flow $U\mathbf{i}$ satisfies

$$\left(1 - \frac{U^2}{c^2}\right)\frac{\partial^2\phi}{\partial x^2} + \frac{\partial^2\phi}{\partial y^2} = \frac{2U}{c^2}\frac{\partial^2\phi}{\partial x \partial t} + \frac{1}{c^2}\frac{\partial^2\phi}{\partial t^2},$$

where c is the speed of sound in the undisturbed gas. The gas is in $y > 0$ and flows past an elastic membrane under tension T which originally lies on $y = 0$ and is subject to small displacements $y = \eta(x,t)$. The boundary conditions on the membrane are that, on $y = 0$,

$$\frac{\partial\phi}{\partial y} = \frac{\partial\eta}{\partial t} + U\frac{\partial\eta}{\partial x}$$

and

$$\frac{p}{\rho_0} = \frac{T}{\rho_0}\left(\frac{\partial^2\eta}{\partial x^2} - \frac{1}{c_m^2}\frac{\partial^2\eta}{\partial t^2}\right) = -\left(\frac{\partial\phi}{\partial t} + U\frac{\partial\phi}{\partial x}\right),$$

where c_m^2 is the speed of sound in the membrane, and ρ_0 is the ambient density in the gas.

By writing

$$\phi = \text{Rl}\,(Ae^{i(kx-\omega t)-\lambda y}), \quad \eta = \text{Rl}\,(ae^{i(kx-\omega t)})$$

with $\text{Rl}\,\lambda \geq 0$, deduce that

$$k^2 - \lambda^2 = \left(\frac{Uk}{c} - \frac{\omega}{c}\right)^2 \quad \text{and} \quad \frac{\lambda T}{\rho_0}\left(k^2 - \frac{\omega^2}{c_m^2}\right) = (Uk - \omega)^2.$$

Deduce that, as $\rho_0 \to 0$ with k real, ω is complex and the motion is unstable if $U > c + c_m$.

This is an example of "flutter", a phenomenon from which aerodynamic surfaces can suffer.

*4.9 (i) Helmholtz' equation in cylindrical polar coordinates (r, θ) is

$$\frac{\partial^2 \Phi}{\partial r^2} + \frac{1}{r}\frac{\partial \Phi}{\partial r} + \frac{1}{r^2}\frac{\partial^2 \Phi}{\partial \theta^2} + k^2 \Phi = 0.$$

To study the far field, write $r = R/\varepsilon$, where $\varepsilon \ll 1$, and use the ansatz $\Phi = Ae^{iku/\varepsilon}$ to show that

$$\left(\frac{\partial u}{\partial R}\right)^2 = 1$$

and

$$2\frac{\partial A}{\partial R} + \frac{A}{R} = 0.$$

Deduce that when waves propagate outwards radially for large r,

$$\Phi \sim \frac{A_0(\theta)}{r^{1/2}}e^{ikr},$$

and hence that $r^{1/2}(\partial \Phi/\partial r - ik\Phi) \to 0$ as $r \to \infty$.

(ii) Helmholtz' equation in spherical polar coordinates (r, θ, ψ) is

$$\frac{\partial^2 \Phi}{\partial r^2} + \frac{2}{r}\frac{\partial \Phi}{\partial r} + \frac{1}{r^2 \sin\theta}\frac{\partial}{\partial \theta}\left(\sin\theta\frac{\partial \Phi}{\partial \theta}\right) + \frac{1}{r^2 \sin\theta}\frac{\partial^2 \Phi}{\partial \psi^2} + k^2 \Phi = 0.$$

Repeat the scaling in (i) to show that, for large r,

$$\Phi \sim \frac{A_0(\theta, \psi)}{r} e^{ikr}$$

and hence that $r(\partial\Phi/\partial r - ik\Phi) \to 0$ as $r \to \infty$.

R4.10 In the frequency domain, ϕ_1 and ϕ_2 satisfy

$$(\nabla^2 + k_1^2)\phi_1 = 0 \text{ in } y > 0$$

and

$$(\nabla^2 + k_2^2)\phi_2 = 0 \text{ in } y < 0,$$

and a class of refraction problems lead to the conditions that $k^2\phi$ and $\partial\phi/\partial y$ are continuous at $y = 0$. Show that if $y = 0$ is irradiated from above by a plane wave

$$\phi_i = Ae^{ik_1(y\cos\theta_1 - x\sin\theta_1)},$$

so that the incident rays are in the $(-\sin\theta_1, \cos\theta_1)$ direction, then

$$\phi_1 = \phi_i + Re^{-ik_1(y\cos\theta_1 + x\sin\theta_1)}$$

and

$$\phi_2 = Te^{ik_2(y\cos\theta_2 - x\sin\theta_2)},$$

where $k_1\sin\theta_1 = k_2\sin\theta_2$.

As long as $\lambda = k_1\sin\theta_1/k_2 < 1$, this "conservation of tangential wave number" is Snell's law of refraction. Since $k_i = \omega/c_i$, it implies that $\sin\theta_1/\sin\theta_2 = c_1/c_2$. The same law holds for optics (see Billingham and King [5] for an analysis of Maxwell's equations), and since the speed of light in water is less than the speed of light in air, Snell's law shows that light is bent towards the normal on entering water, making ponds seem shallower than they really are.

If $k_1 > k_2$, it is possible for λ to be greater than unity and in this case there is total internal reflection. Show that if $\lambda > 1$,

$$\phi_2 = Be^{k_2\sqrt{\lambda^2 - 1}y - i\lambda k_2 x}$$

and that, in this case, $|R| = |A|$.

4.11 Verify that the longitudinal elastic wave

$$\mathbf{X} = \mathrm{Rl}\begin{pmatrix} \cos\alpha \\ \sin\alpha \end{pmatrix} \exp\left[i\left(\frac{\omega}{c_p}(x\cos\alpha + y\sin\alpha) - \omega t \right) \right]$$

satisfies (3.72) and is the gradient of a potential that satisfies (3.75). Show that if this wave is incident from the region $x < 0$ onto a rigid boundary at $x = 0$ (at which $\mathbf{X} = \mathbf{0}$) then the reflected wave is

$$\mathbf{X}_r = \mathbf{X}_{rp} + \mathbf{X}_{rs},$$

where

$$\mathbf{X}_{rp} = Rlr_p \begin{pmatrix} -\cos\gamma \\ \sin\gamma \end{pmatrix} \exp\left[i\left(\frac{\omega}{c_p}(-x\cos\gamma + y\sin\gamma) - \omega t\right)\right]$$

and

$$\mathbf{X}_{rs} = Rlr_s \begin{pmatrix} \sin\beta \\ \cos\beta \end{pmatrix} \exp\left[i\left(\frac{\omega}{c_s}(-x\cos\beta + y\sin\beta) - \omega t\right)\right]$$

both satisfy (3.72). Show that $\mathbf{X}_{rs}$ is the curl of a vector whose components satisfy (3.76) and represents a transverse wave travelling with speed c_s. Show also that $\gamma = \alpha$ and $\sin\beta = (c_s/c_p)\sin\alpha$, and that

$$r_p = \frac{\cos(\alpha + \beta)}{\cos(\alpha - \beta)}, \quad r_s = -\frac{\sin 2\alpha}{\cos(\alpha - \beta)}.$$

4.12 Elastic waves in a plate can be modelled by the equation

$$D\left(\frac{\partial^2}{\partial x^2} + \frac{\partial^2}{\partial y^2}\right)^2 u + \rho\frac{\partial^2 u}{\partial t^2} = 0,$$

where $u(x, y)$ is the normal displacement of the plate and D is a positive constant proportional to the Young's modulus of the material from which the plate is made. Show that the dispersion relation is

$$\rho\omega^2 = D(k_1^2 + k_2^2)^2.$$

In one dimension, the model reduces to the *beam equation*

$$D\frac{\partial^4 u}{\partial x^4} + \rho\frac{\partial^2 u}{\partial t^2} = 0$$

and suitable boundary conditions for a beam which is clamped at both ends are $u = \partial u/\partial x = 0$ at the ends. Show that the resonant frequencies for a clamped beam are not rationally related to each other.

Note that this result is in contrast to solutions for the transverse vibrations of a string for which

$$-T\frac{\partial^2 u}{\partial x^2} + \rho\frac{\partial^2 u}{\partial t^2} = 0$$

with $u = 0$ at $x = 0, l$, and the resonant frequencies are $(n\pi/l)\sqrt{(T/\rho)}$. Note also that the real and imaginary parts of the solution of (4.86) satisfy a beam equation.

*4.13 (i) The frequency domain model for waves in a periodic medium in which the inhomogeneity is weak is the *Mathieu equation*

$$\frac{d^2\Phi}{dx^2} + (k^2 + \varepsilon \cos 2x)\Phi = 0,$$

where $k = O(1)$ and $\varepsilon \ll 1$. Show that a perturbation solution[23]

$$\Phi \sim A \cos kx + B \sin kx + \varepsilon\phi_1 + \varepsilon^2\phi_2 + \cdots$$

reveals that the terms $\varepsilon\phi_1, \varepsilon^2\phi_2, \varepsilon^3\phi_3 \ldots$ cannot all remain small compared to the lowest order term for all x if k is an integer. Suppose that $k^2 = 1 + \kappa\varepsilon$, where κ is $O(1)$. Show that

$$\frac{d^2\phi_1}{dx^2} + \phi_1 = -\kappa(A \cos x + B \sin x)$$
$$- \frac{1}{2}(A(\cos 3x + \cos x) + B(\sin 3x - \sin x)).$$

Deduce that Φ can only be periodic to order ε if $\kappa = \frac{1}{2}$ and $A = 0$, or if $\kappa = -\frac{1}{2}$ and $B = 0$.

(ii) If $k^2 = 1 + \kappa\varepsilon$ as above, but x is large so that $x = X/\varepsilon$, where $X = O(1)$, show that the WKB solution of the equation is

$$\Phi \sim Ae^{iX/\varepsilon} + A^*e^{-iX/\varepsilon},$$

where

$$2i\frac{dA}{dX} + \kappa A + \frac{1}{2}A^* = 0 \text{ and } -2i\frac{dA^*}{dX} + \kappa A^* + \frac{1}{2}A = 0.$$

Deduce that A and A^* grow or decay exponentially in X if $|\kappa| < \frac{1}{2}$ and that they are oscillatory in X if $|\kappa| > \frac{1}{2}$.

This example shows that waves in such a periodic medium will decay if k is in the *stop band* $-\varepsilon/4 < k - 1 < \varepsilon/4$. It can be shown that similar stop bands exist near any integer value of k (including 0) as shown in Figure 4.5. Hence, over large regions of the (k, ε) parameter space, the material acts to damp waves exponentially rather than allowing them to propagate. This is an example of *Floquet Theory*, which, in higher dimensions, is associated with the names of Bloch and Brillouin.

---

[23] This is an example of the *Born approximation* in which the solution of Helmholtz' equation in a medium that is only weakly inhomogeneous is expanded in such a power series.

4.14 A one-dimensional acoustic resonator is closed at $x = 0$ and is driven at $x = 1$ by a piston which oscillates with a frequency which is much lower than any of the resonant frequencies of the pipe. Show that a suitable nondimensional model is

$$\frac{\partial^2 \phi}{\partial x^2} = \frac{\partial^2 \phi}{\partial t^2} \quad \text{for } 0 < x < 1,$$

with $\partial \phi / \partial x = 0$ on $x = 0$ and $\partial \phi / \partial x = \sin \varepsilon t$ at $x = 1$, where $\varepsilon \ll 1$. Assuming a periodic response, solve this problem and show that as $\varepsilon \to 0$, the pressure response has amplitude of $O(\varepsilon^{-1})$ but that the gas velocity will never be greater than its maximum value at the piston.

4.15 (i) Suppose that N is constant in the inertial wave model (3.26) and that $w = \text{Rl}\,(W(x, y, z)e^{-i\omega t})$. Show that if g is sufficiently large, then

$$(N^2 - \omega^2)\left(\frac{\partial^2 W}{\partial x^2} + \frac{\partial^2 W}{\partial y^2}\right) = \omega^2 \frac{\partial^2 W}{\partial z^2},$$

and deduce that waves can radiate to infinity if $\omega^2 < N^2$.

N is called the *Brunt-Väisälä frequency*.

(ii) Show from (3.32) that if $p' = P(x, y, z)e^{-i\omega t}$ in steady rotating flow at high Rossby number, then P satisfies

$$\frac{\partial^2 P}{\partial x^2} + \frac{\partial^2 P}{\partial y^2} + \frac{\partial^2 P}{\partial z^2} = \frac{4\Omega^2}{\omega^2} \frac{\partial^2 P}{\partial z^2}.$$

Show that waves can radiate to infinity if $\omega^2 < 4\Omega^2$.

Note that in both these cases, there is a *cut-off frequency* above which waves cannot radiate to infinity.

4.16 Show that the wave crests and troughs in the far field ship wave pattern (4.56) are given by the curves $k_1 x + k_2 y = c$, where k_1, k_2 are related by (4.57) and c is a constant. Remembering that $y = \lambda x$, where λ is given by (4.59) and putting $U^4/g^2 = 1$ for simplicity, show that the crests and troughs are given parametrically in terms of k_1 by

$$x = \frac{c(2k_1^2 - 1)}{k_1^3}, \quad y = \mp \frac{(k_1^2 - 1)^{1/2} c}{k_1^3}.$$

Show that, for a fixed c, y is maximum when $k_1^2 = 3/2$ and $y/x = \mp 1/2\sqrt{2}$. Show that $dy/dx = \mp 1/(k_1^2 - 1)^{1/2}$ and sketch these curves for fixed c. Hence show that the crests in a ship's wake consist of two families of curves as shown in Figure 4.6.

*4.17 Steady two-dimensional flow in a slender nozzle is modelled by (2.6), (2.7) and (2.20) in the form

$$\frac{\partial}{\partial x}(\rho u) + \frac{\partial}{\partial y}(\rho v) = 0$$

$$\rho\left(u\frac{\partial u}{\partial x} + v\frac{\partial u}{\partial y}\right) = -\frac{\partial p}{\partial x},$$

$$\rho\left(u\frac{\partial v}{\partial x} + v\frac{\partial v}{\partial y}\right) = -\frac{\partial p}{\partial y},$$

and

$$\left(u\frac{\partial}{\partial x} + v\frac{\partial}{\partial y}\right)\left(\frac{p}{p^\gamma}\right) = 0$$

with

$$\frac{v}{u} = \pm\frac{\varepsilon}{2}\frac{d\bar{A}}{dx} \quad \text{on } y = \pm\frac{1}{2}(A_0 + \varepsilon\bar{A}(x)),$$

where $A_0 = $ constant. Show that if $\varepsilon x = X$, $u = u_0 + \varepsilon u'$, $v = \varepsilon^2 v'$, $p = p_0 + \varepsilon p'$ and $\rho = \rho_0 + \varepsilon\rho'$, where u_0, p_0 and ρ_0 are constant, then, to lowest order in ε,

$$\rho_0\frac{\partial u'}{\partial X} + u_0\frac{\partial \rho'}{\partial X} + \rho_0\frac{\partial v'}{\partial y} = 0,$$

$$\rho_0 u_0\frac{\partial u'}{\partial X} = -\frac{\partial p'}{\partial X}, \quad 0 = \frac{\partial p'}{\partial y}$$

and

$$p' = \frac{\gamma p_0}{\rho_0}\rho',$$

with

$$v' = \pm\frac{u_0}{2}\frac{d\bar{A}}{dX} \quad \text{on } y = \pm\frac{A_0}{2}.$$

Deduce that, if $\bar{u} = 1/A_0 \int_{-A_0/2}^{A_0/2} u'\,dy$, and similarly for $\bar{p}, \bar{\rho}$, then

$$A_0\rho_0\frac{d\bar{u}}{dx} + A_0 u_0\frac{d\bar{\rho}}{dx} + \rho_0 u_0\frac{d\bar{A}}{dx} = 0,$$

$$\rho_0 u_0\frac{d\bar{u}}{dx} + \frac{d\bar{p}}{dx} = 0$$

and

$$\bar{p} = \frac{\gamma p_0}{\rho_0}\bar{\rho}.$$

Hence, show that $\bar{\rho} = \rho_0\bar{A}M^2/A_0(1 - M^2)$, where $M^2 = u_0^2\rho_0/\gamma p_0$.

*4.18 (i) Show that if $\phi(x, y)$ is given by (4.67), then

$$\frac{\partial \phi}{\partial y} = 2\beta^2 y \int_0^l \frac{g(\xi)\,d\xi}{(x - \xi)^2 + \beta^2 y^2}.$$

Noting that, as $y \to 0$, the major contribution to the integral comes from near $\xi = x$, show that

$$\lim_{y \downarrow 0} \frac{\partial \phi}{\partial y} = \begin{cases} 2\pi\beta g(x), & 0 < x < l \\ 0, & \text{otherwise} \end{cases}$$

and

$$\lim_{y \uparrow 0} \frac{\partial \phi}{\partial y} = \begin{cases} -2\pi\beta g(x), & 0 < x < l \\ 0, & \text{otherwise.} \end{cases}$$

Use the same type of argument to show that the integral in (4.75) tends to $h(x)[\sinh^{-1}((l - x)/\beta r) + \sinh^{-1} x/\beta r]$ as $r \to 0$ and deduce that $\phi \sim -2h(x) \log r$ in this limit.

(ii) Show that the potential

$$\phi(x, y) = \int_0^l h(\xi) \tan^{-1}\left(\frac{x - \xi}{\beta y}\right) d\xi \qquad (*)$$

satisfies (4.63), where the function $\tan^{-1}$ is defined to lie between $-\pi/2$ and $\pi/2$. Show that as $y \downarrow 0$,

$$\tan^{-1}\left(\frac{x - \xi}{\beta y}\right) \sim \pm\frac{\pi}{2} - \frac{\beta y}{x - \xi} + O(y^2) \qquad (**)$$

according to whether $x - \xi > 0$ or $x - \xi < 0$, respectively. Show directly from (*) that, for $0 < x < l$ with y fixed,

$$\phi(x, y) = \lim_{\varepsilon \to 0} \left(\int_0^{x-\varepsilon} + \int_{x+\varepsilon}^l\right) h(\xi) \tan^{-1}\left(\frac{x - \xi}{\beta y}\right) d\xi.$$

Hence, use (**) to show that, for small positive values of y, ϕ is approximately

$$\int_0^{x-\varepsilon} h(\xi) \left(\frac{\pi}{2} - \frac{\beta y}{x - \xi}\right) d\xi + \int_{x+\varepsilon}^l h(\xi) \left(-\frac{\pi}{2} - \frac{\beta y}{x - \xi}\right) d\xi.$$

Deduce that, as $y \downarrow 0$,

$$\frac{\partial \phi}{\partial y} \to -\beta \int_0^l \frac{h(\xi)}{x - \xi} d\xi,$$

where the *Cauchy Principal Value* integral is defined by

$$\fint_0^l \frac{h(\xi)}{x-\xi}d\xi = \lim_{\varepsilon \to 0}\left(\int_0^{x-\varepsilon} + \int_{x+\varepsilon}^l\right)\frac{h(\xi)}{x-\xi}d\xi.$$

Show that $\partial\phi/\partial y$ takes the same value as $y \uparrow 0$ and deduce that the problem for subsonic flow past an infinitely thin *asymmetric* aerofoil $y = \varepsilon f_A(x)$ requires us to solve the *singular integral equation*

$$-\frac{U}{\beta}f_A'(x) = \fint_0^l \frac{h(\xi)\,d\xi}{x-\xi}.$$

Note that an arbitrary wing shape $\varepsilon f_\pm(x)$ can always be written as $\varepsilon(\pm f_S + f_A)$, where $f_S = \frac{1}{2}(f_+ - f_-), f_A = \frac{1}{2}(f_+ + f_-)$.

4.19 Solve the problem of subsonic flow past a symmetric thin wing by using Fourier transforms as follows. If $\bar\phi$ is defined by $\bar\phi = \int_{-\infty}^{\infty}\phi(x,y)e^{ikx}dx$, show that (4.66) and (4.65) lead to the problem

$$\frac{d^2\bar\phi}{dy^2} - \beta^2 k^2\bar\phi = 0 \text{ in } y > 0,$$

with

$$\frac{d\bar\phi}{dy} = U\bar F(k) \text{ on } y = 0,$$

where $\bar F(k)$ is the Fourier transform of $f'(x)$, which is defined to be zero for $x < 0, x > l$. Hence, show that

$$\frac{d\bar\phi}{dy} = U\bar F(k)e^{-\beta|k|y}.$$

From Exercise 4.4, the Fourier transform of $\beta y/(x^2 + \beta^2 y^2)$ is $\pi e^{-\beta|k|y}$; use this result to determine $\partial\phi/\partial y$ and hence show that

$$\phi(x,y) = \frac{U}{2\pi\beta}\int_{-\infty}^{\infty} f'(x)\log((x-\xi)^2 + \beta^2 y^2)\,d\xi.$$

*4.20 (i) Solve the problem of subsonic flow past a slender axisymmetric body at zero incidence by using Fourier transforms as follows. If $\bar\phi$ is defined as

$$\bar\phi(r,k) = \int_{-\infty}^{\infty}\phi(x,r)e^{ikx}\,dx,$$

where the velocity potential of the flow is $Ux + \varepsilon\phi$, show that $\bar\phi$ satisfies the equation

$$\frac{d^2\bar\phi}{dr^2} + \frac{1}{r}\frac{d\bar\phi}{dr} - \beta^2 k^2\bar\phi = 0.$$

The solution of this equation which tends to zero as $r \to \infty$ is the Bessel function $K_0(\beta|k|r)$. By writing

$$\bar{\phi} = \bar{A}(k)K_0(\beta|k|r)$$

and using the fact that $K_0(\beta|k|r) \sim -\log r$ as $r \to 0$ in the boundary condition (4.74), show that $\bar{A}(k)$ is the Fourier transform of $A(x)$, where

$$A(x) = -\varepsilon U R(x)R'(x) \quad \text{for } 0 < x < l.$$

Given that $K_0(\beta|k|r)$ is the Fourier transform of $\dfrac{1}{2\sqrt{x^2 + \beta^2 r^2}}$, show that

$$\phi(x, r) = \int_0^l \frac{-\varepsilon U R(\xi)R'(\xi)}{2((x-\xi)^2 + \beta^2 r^2)^{1/2}} d\xi.$$

(ii) A slender body $r = \varepsilon R(x)$ is now placed in a subsonic stream which makes a small angle α to the axis of the body. If the free stream velocity is

$$(U\alpha \cos \theta, -U\alpha \sin \theta, U)$$

in the (r, θ, x) directions in cylindrical polar coordinates, show that the boundary condition on the body is

$$\frac{\partial \phi}{\partial r} = UR'(x) - \frac{U\alpha}{\varepsilon} \cos \theta. \tag{*}$$

Assuming that α/ε is $O(1)$ and using the Fourier transform, as in part (i), show that

$$\bar{\phi} = \bar{A}(k)K_0(\beta|k|r) + \bar{B}(k)\frac{\partial}{\partial r}(K_0(\beta|k|r)) \cos \theta.$$

If $B(x)$ is the inverse transform of $\bar{B}(k)$, show that

$$B(x) = -U\alpha\varepsilon(R(x))^2,$$

and deduce that the full solution is

$$\phi(r, \theta, x) = -\frac{\varepsilon U}{2} \int_0^l \frac{R(\xi)R'(\xi)\, d\xi}{((x-\xi)^2 + \beta^2 r^2)^{1/2}}$$
$$-\frac{U\alpha\varepsilon \cos \theta}{2} \frac{\partial}{\partial r} \int_0^l \frac{(R(\xi))^2\, d\xi}{((x-\xi)^2 + \beta^2 r^2)^{1/2}}.$$

R 4.21 A thin wing is placed at a small angle of incidence α in a steady supersonic stream, so that the wing is given by

$$y = \varepsilon f_\pm(x) - \alpha x \text{ for } 0 < x < l,$$

where $\alpha = O(\varepsilon)$. Show that the drag is

$$\frac{\rho_0 U^2}{B} \int_0^l [(\varepsilon f_+'(x) - \alpha)^2 + (\varepsilon f_-'(x) - \alpha)^2] \, dx.$$

Hence, confirm that the drag on a flat plate of length l at a small angle of incidence α is $(2\rho_0 U^2/B)\alpha^2 l$.

Show that if $\alpha = 0$ and the wing has a cross-section

$$f_+ = -f_- = \begin{cases} mx, & 0 < x < \dfrac{h}{m} \\[2mm] \dfrac{h(l-x)}{l-h/m}, & \dfrac{h}{m} < x < l \end{cases}$$

with a given thickness h, then the drag is minimised for a diamond shape with $h/m = l/2$.

4.22 A plane high-frequency wave, given by $\Phi = e^{-ikx}$, is incident from the right on a parabolic reflector $y^2 = 4x$. If the reflected wave is given by $\Phi = Ae^{iku}$, show that

$$x = \tan^2 \frac{s}{2}, \quad y = 2\tan \frac{s}{2}, \quad u = -\tan^2 \frac{s}{2}, \quad (-\pi < s < \pi)$$

is suitable boundary data for u. By showing that $p = \cos s$ and $q = -\sin s$ and solving Charpit's equations (4.84), show that the reflected rays are given by

$$x = 2\tau \cos s + \tan^2 \frac{s}{2}, \quad y = -2\tau \sin s + 2\tan \frac{s}{2}, \quad u = 2\tau - \tan^2 \frac{s}{2}.$$

Deduce that all the reflected rays pass through the focus $(1,0)$.

4.23 By writing $A = F(Y/\sqrt{x})$ in (4.86) for $x > 0$, show that

$$F(\eta) = c_1 \int_\eta^\infty e^{i\tau^2/2} d\tau + c_2,$$

where c_1 and c_2 are constants. (The integral is called a Fresnel integral). Deduce that if the region $Y > 0$ is in shadow so that $A \to 0$ as $Y \to \infty$ and $Y < 0$ is illuminated so that $A \to 1$ as $Y \to -\infty$, then

$$A = \frac{e^{i\pi/4}}{\sqrt{2\pi}} \int_{Y/\sqrt{x}}^\infty e^{i\tau^2/2} d\tau.$$

By integrating by parts, show that to lowest order the real and imaginary parts of A oscillate as $Y \to \infty$. [Note that $\int_{-\infty}^{\infty} e^{it^2} dt = \sqrt{\pi} e^{i\pi/4}$.]

4.24 Suppose waves are confined to a thin layer inside a circular boundary $r = 1$ in plane polar coordinates. Show that, if we write $\Phi = A(r)^{ik\theta}$ in (4.80), then

$$\frac{d^2 A}{dr^2} + \frac{1}{r}\frac{dA}{dr} + k^2\left(1 - \frac{1}{r^2}\right)A = 0.$$

Hence, show that a paraxial approximation, where $k \gg 1$ and $r = 1 - k^{-\alpha}\rho$ with $\alpha > 0$, leads to

$$k^{2\alpha}\frac{d^2 A}{d\rho^2} - \frac{k^\alpha}{1 - k^{-\alpha}\rho}\frac{dA}{d\rho} + k^2\left(-2k^{-\alpha}\rho + O(k^{-2\alpha})\right)A = 0$$

Deduce that the appropriate choice for α is $\alpha = \frac{2}{3}$ and then, as $k \to \infty$, the lowest order approximation for A satisfies

$$\frac{d^2 A}{d\rho^2} - 2\rho A = 0.$$

This is a version of Airy's equation (see Exercise 4.1) and it applies locally to any thin layer when the radius of curvature of the boundary is much greater than the wavelength. Note that the layer is of width $O(k^{-2/3})$ which is thinner than a shadow boundary. This solution not only describes caustics, where $A \to 0$ exponentially as $\rho \to \infty$ and algebraically as $\rho \to -\infty$, but also "whispering gallery" waves.

4.25 Suppose that in (4.89), the function f is spherically symmetric so that $f(x, y, z) = F(r)$, where $x^2 + y^2 + z^2 = r^2$. Show that, at any point $(0, 0, z)$,

$$\phi(0, 0, z, t) = \frac{ct}{4\pi}\int_0^{2\pi}\int_0^\pi F((z^2 + 2zct\cos\theta + c^2t^2)^{1/2})\sin\theta\, d\theta\, d\phi$$

$$= \frac{1}{2z}[g(z + ct) - g(|z - ct|)],$$

where $g'(r) = rF(r)$. By rotating the axes, deduce that

$$\phi(x, y, z, t) = \frac{1}{2r}[g(r + ct) - g(|r - ct|)].$$

Verify that this gives a spherically symmetric solution of the wave equation which satisfies $\phi = 0$ and $\partial\phi/\partial t = F(r)$ at $t = 0$.

4.26 Suppose that, in (4.89), the function $f(x, y, z)$ is independent of z. Write $\rho = ct \sin \theta$ when $|\rho| < ct$ and show that

$$\phi(x, y, t) = \frac{1}{2\pi} \int_0^{2\pi} \int_0^{ct} f(x + \rho \cos \phi, y + \rho \sin \phi) \frac{\rho \, d\rho \, d\phi}{\sqrt{c^2 t^2 - \rho^2}}$$

$$= \frac{1}{2\pi} \int \int_S \frac{f(\xi, \eta) \, d\xi \, d\eta}{(c^2 t^2 - (x - \xi)^2 - (y - \eta)^2)^{1/2}},$$

where S is the interior of the sphere $(x - \xi)^2 + (y - \eta)^2 = c^2 t^2$. Hence, show that there is no sharp termination of the wave in the case when $f(\xi, \eta)$ vanishes outside some bounded region in the (ξ, η) plane.

Show also that if $f(\xi, \eta)$ is localised near $\xi = 0$, $\eta = 0$, we retrieve the solution (4.88) where for some constant λ,

$$\phi = \begin{cases} \dfrac{\lambda}{(c^2 t^2 - r^2)^{1/2}}, & r < ct \\ 0, & r > ct, \end{cases}$$

where $r^2 = x^2 + y^2$.

*4.27 (i) The spherically symmetric wave equation in n dimensions is

$$\frac{\partial^2 \phi}{\partial r^2} + \frac{n - 1}{r} \frac{\partial \phi}{\partial r} = \frac{1}{c^2} \frac{\partial^2 \phi}{\partial t^2}.$$

Show that when n is odd, the general solution is

$$\left(\frac{1}{r} \frac{\partial}{\partial r} \right)^{(n-3)/2} \frac{f(ct \pm r)}{r}.$$

(ii) From (4.78), show that the general solution for outward propagating axisymmetric waves in two dimensions is

$$\phi = \int_r^{ct} \frac{f(ct - s) \, ds}{\sqrt{s^2 - r^2}},$$

where $m(z/c) = f(z)$.

It can be shown that in an even number of dimensions n, the corresponding solution is

$$\left(\frac{1}{r} \frac{\partial}{\partial r} \right)^{(n-2)/2} \left[\int_r^{ct} \frac{f(ct - s) \, ds}{\sqrt{s^2 - r^2}} \right],$$

which is the same as the result of (i) in terms of pseudo-differential operators.

4.28 By taking the integral in (4.75) from $x - i\beta r$ to $x + i\beta r$ and assuming that $h(\xi)$ is analytic throughout, show that

$$\phi(x, r) = \int_{x-i\beta r}^{x+i\beta r} \frac{h(\xi)\,d\xi}{((x-\xi)^2 + \beta^2 r^2)^{1/2}}$$

$$= \int_0^\pi ih(x + i\beta r \cos\theta)\,d\theta$$

satisfies the axisymmetric equation (4.73). Verify that ϕ is analytic on $r = 0$ and that

$$\phi(x, r) = \frac{1}{\pi}\int_0^\pi \phi(x + i\beta r \cos\theta, 0)\,d\theta. \tag{†}$$

Note the difference between this result, which holds for axisymmetric potentials that are analytic at $r = 0$ for all x, and the representation (4.75) where $\phi \sim -2h(x)\log r$ as $r \to 0$.

4.29 Show that

$$\phi(x, r) = \int_0^{x-Br} \frac{m(\xi)\,d\xi}{\sqrt{(x-\xi)^2 - B^2 r^2}} = \int_0^{\cosh^{-1}(x/Br)} m(x - Br\cosh t)\,dt$$

and hence confirm that $\dfrac{\partial^2\phi}{\partial r^2} + \dfrac{1}{r}\dfrac{\partial\phi}{\partial r} - B^2\dfrac{\partial^2\phi}{\partial x^2} = 0.$

Chapter 5
Nonlinear Waves in Fluids

5.1 Introduction

We have already encountered several deficiencies in the theories presented in Chapter 4 that indicate the limitations of the linear approximation. In particular, we have seen that the linear theory cannot deal with the following situations:

(i) The transition from subsonic to supersonic flow past a thin wing.
(ii) The strange behaviour in a converging-diverging nozzle when sonic conditions are attained.
(iii) The response of a system near resonance.

More generally, we may also ask what happens when bodies which are thick or "blunt-nosed" are placed in a compressible stream, or when the elevation of a surface gravity wave is comparable to the depth of the water, or when the amplitude of the motion of a gas in a resonator is comparable to the length scale of the system.

In this chapter, we will consider three specific nonlinear models, namely, unsteady one-dimensional gas dynamics, two-dimensional steady gas dynamics and shallow water theory. However, before we embark on these models, we will discuss a simple paradigm example of a nonlinear system that will help us to understand the more complicated systems that will follow.

We consider the equation

$$\frac{\partial u}{\partial t} + u \frac{\partial u}{\partial x} = 0, \tag{5.1}$$

© Springer Science+Business Media New York 2015
H. Ockendon, J.R. Ockendon, *Waves and Compressible Flow*,
Texts in Applied Mathematics 47, DOI 10.1007/978-1-4939-3381-5_5

defined for $-\infty < x < \infty$ and $t > 0$, and with $u(x, 0) = u_0(x)$ prescribed.[1] This nonlinear equation can be solved exactly. The general solution is $u = F(x - ut)$ for any function F and so

$$u = u_0(x - ut) \tag{5.2}$$

describes the solution implicitly. An alternative approach is to note that, along the characteristics of (5.1),

$$\frac{dx}{dt} = u, \quad \frac{du}{dt} = 0,$$

and so u is constant on a characteristic, which is therefore a straight line. Thus, the characteristics can be drawn just by using the initial slopes, as given by $u_0(x)$, and an example is shown in Figure 5.1.

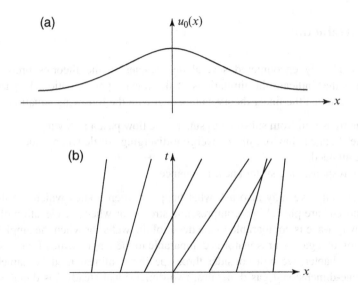

Fig. 5.1 (a) Initial data for equation (5.1). (b) Characteristics of equation (5.1).

We are immediately confronted by one of the fundamental difficulties of nonlinear hyperbolic partial differential equations, which is that, since the slope of the characteristic depends on the solution, it is possible for characteristics to *intersect*, leading to a multivalued solution for u.

If $u_0(x)$ is smooth and has finite slope everywhere, then the solution (5.2) will hold for small values of t but, except in rare situations such as when u_0 is monotonic

---

[1] This is an example of what is often known as the *kinematic wave equation*, namely,

$$\frac{\partial u}{\partial t} + \frac{\partial}{\partial x}(f(u)) = 0$$

for some function f.

increasing for all x, u will eventually become multivalued as a result of intersecting characteristics. Of course, it may be possible to allow multivalued solutions if u is the profile of, say, a water wave, but, in general, multivalued solutions are not physically acceptable; the remedy for this situation is the introduction of a discontinuity or shock wave as will be described in Chapter 6.

5.2 Models for Nonlinear Waves

5.2.1 One-dimensional Unsteady Gas Dynamics

We now consider some exact solutions of the full equations for the flow of a perfect compressible gas. The first case we consider is that of one-dimensional unsteady flow and, in this case, equations (2.6), (2.7) and (2.22) reduce to

$$\frac{\partial \rho}{\partial t} + \frac{\partial}{\partial x}(\rho u) = 0, \tag{5.3}$$

$$\frac{\partial u}{\partial t} + u\frac{\partial u}{\partial x} + \frac{1}{\rho}\frac{\partial p}{\partial x} = 0 \tag{5.4}$$

and

$$\frac{d}{dt}\left(\frac{p}{\rho^\gamma}\right) = 0. \tag{5.5}$$

We assume that the flow is homentropic, as it would be if the gas was initially in a uniform state. This assumption means that, from (5.5), p/ρ^γ is constant and since $c^2 = dp/d\rho$, we can deduce that

$$\frac{dp}{p} = \gamma\frac{d\rho}{\rho} = \frac{2\gamma dc}{(\gamma - 1)c}. \tag{5.6}$$

Using (5.6), (5.3) and (5.4) can be written in terms of u and c alone as

$$\frac{2}{\gamma - 1}\frac{\partial c}{\partial t} + \frac{2u}{\gamma - 1}\frac{\partial c}{\partial x} + c\frac{\partial u}{\partial x} = 0 \tag{5.7}$$

and

$$\frac{\partial u}{\partial t} + u\frac{\partial u}{\partial x} + \frac{2c}{\gamma - 1}\frac{\partial c}{\partial x} = 0. \tag{5.8}$$

Adding and subtracting (5.7) and (5.8) leads to

$$\left(\frac{\partial}{\partial t} + (u \pm c)\frac{\partial}{\partial x}\right)\left(u \pm \frac{2c}{\gamma - 1}\right) = 0, \tag{5.9}$$

and hence, we can see that

$$u \pm \frac{2c}{\gamma - 1} \text{ is constant on } \frac{dx}{dt} = u \pm c. \tag{5.10}$$

The quantities $u \pm 2c/(\gamma - 1)$ are the Riemann invariants, and the curves $dx/dt = u \pm c$ are the characteristics[2] of the second-order hyperbolic system of equations (5.9). More generally, for non-homentropic flow, the three equations (5.3)–(5.5) form a *third-order hyperbolic system* with characteristics given by $dx/dt = u \pm c$ and $dx/dt = u$, where $c^2 = \gamma p/\rho$. The "third" characteristic is the particle path. We will define the characteristics of a system of equations more precisely in Section 5.2.3.

5.2.2 Two-dimensional Steady Homentropic Gas Dynamics

Our second example is two-dimensional steady compressible flow. We again assume that the flow is homentropic so that p/ρ^γ is constant, and then, (2.6) and (2.7) can be written as

$$\frac{\partial}{\partial x}(\rho u) + \frac{\partial}{\partial y}(\rho v) = 0, \tag{5.11}$$

$$u\frac{\partial u}{\partial x} + v\frac{\partial u}{\partial y} + \frac{c^2}{\rho}\frac{\partial \rho}{\partial x} = 0 \tag{5.12}$$

and

$$u\frac{\partial v}{\partial x} + v\frac{\partial v}{\partial y} + \frac{c^2}{\rho}\frac{\partial \rho}{\partial y} = 0. \tag{5.13}$$

We use the same ideas to simplify these equations as we did in the unsteady one-dimensional case, but, because there are three equations now rather than two, the technical details are more complicated. Nevertheless, we can easily eliminate $\partial \rho/\partial x$ and $\partial \rho/\partial y$ from (5.11)–(5.13) to obtain

$$(c^2 - u^2)\frac{\partial u}{\partial x} - uv\left(\frac{\partial u}{\partial y} + \frac{\partial v}{\partial x}\right) + (c^2 - v^2)\frac{\partial v}{\partial y} = 0. \tag{5.14}$$

To get a closed system for u and v, we can note that the flow will be irrotational by Crocco's Theorem (Exercise 2.5); hence, c^2 is given from Bernoulli's equation (2.30) as

$$c^2 = c_0^2 - \frac{\gamma - 1}{2}(u^2 + v^2), \tag{5.15}$$

---

[2] For brevity, we will sometimes refer to these two families of characteristics as the *positive* and *negative* characteristics, respectively.

where c_0 is the value of c when the flow is brought to rest homentropically, and

$$\frac{\partial u}{\partial y} - \frac{\partial v}{\partial x} = 0. \tag{5.16}$$

We could now define a potential function ϕ and then equations (5.14) and (5.15) yield

$$\left(c^2 - \left(\frac{\partial \phi}{\partial x}\right)^2\right)\frac{\partial^2 \phi}{\partial x^2} - 2\frac{\partial \phi}{\partial x}\frac{\partial \phi}{\partial y}\frac{\partial^2 \phi}{\partial x \partial y} + \left(c^2 - \left(\frac{\partial \phi}{\partial y}\right)^2\right)\frac{\partial^2 \phi}{\partial y^2} = 0, \tag{5.17}$$

where $c^2 = c_0^2 - [(\gamma - 1)/2]|\nabla \phi|^2$. Equation (5.17) is a second-order quasi-linear equation for ϕ to which we will return later, but, for now, we work with the first-order system (5.14) and (5.16) and use the methods of Section 5.2.1 to try to write these equations in a more convenient form. Adding (5.14) to a multiple of (5.16), we find that by choosing the multipliers to be $\pm c\sqrt{u^2 + v^2 - c^2}$, the equations emerge in a form that can be integrated (Exercise 5.2); however, to proceed, we do have to make the all-important assumption that $u^2 + v^2 \geq c^2$. We find that

$$(c^2 - u^2)du - (uv \pm c\sqrt{u^2 + v^2 - c^2})dv = 0 \tag{5.18}$$

on the characteristic curves given by

$$\frac{dy}{dx} = \frac{-uv \pm c\sqrt{u^2 + v^2 - c^2}}{c^2 - u^2}. \tag{5.19}$$

If we introduce the new variables

$$\mu = \sin^{-1}\left(\frac{c}{\sqrt{u^2 + v^2}}\right) \quad \text{and} \quad \theta = \tan^{-1}\left(\frac{v}{u}\right),$$

we find that (5.18) can be integrated to give the Riemann invariants

$$\theta \pm \left[\mu + \frac{1}{\lambda}\tan^{-1}(\lambda \cot \mu)\right], \tag{5.20}$$

where $\lambda^2 = (\gamma - 1)/(\gamma + 1)$; these invariants are constant on the characteristics given by

$$\frac{dy}{dx} = \tan(\theta \mp \mu), \tag{5.21}$$

respectively. The angle μ defined above is called the *Mach angle* and, from (5.21), we see that the characteristics always make an angle $\mp \mu$ with the streamlines, and these will again be called the negative and positive characteristics, respectively. Note that, as can be seen from (5.18) and (5.19), these characteristics and Riemann invariants are real in view of our assumption that $u^2 + v^2 \geq c^2$, so that the flow is supersonic throughout. As in Section 5.2.1, even in non-homentropic flow, equations (5.11)–(5.13) are a hyperbolic system as long as $u^2 + v^2 > c^2$ and the characteristics will be given by (5.19) and $dy/dx = v/u$. The third characteristic is, naturally, the streamline.

5.2.3 Shallow Water Theory

It would be nice if we could treat nonlinear surface gravity waves as described by (3.8), (3.10) and (3.11) similarly, but this system is even more difficult to analyse than the models in Sections 5.2.1 and 5.2.2. One way we can make headway here is if we restrict ourselves to analysing the effect of nonlinearity on water that is *shallow*. What we mean by this is that the mean depth of the water, h, is comparable to the amplitude of the waves but small compared to their wavelength λ.

To see the implications of these assumptions, we nondimensionalise the variables by writing

$$x = \lambda X, \quad z = hZ, \quad u = U\hat{u}, \quad t = \frac{\lambda}{U}T,$$

where U is a typical horizontal velocity; for the moment, we are only considering two-dimensional flows with z measured along the upward vertical. From the continuity equation for an incompressible fluid (2.6), the appropriate nondimensionalisation for the vertical component of velocity is

$$w = \frac{hU}{\lambda}\hat{w},$$

and from the x-component of the momentum equation (2.7), the appropriate scaling for the pressure is

$$p = p_0 + \rho U^2 \hat{p},$$

where p_0 is the pressure in the atmosphere. Now (2.6) and (2.7) become

$$\frac{\partial \hat{u}}{\partial X} + \frac{\partial \hat{w}}{\partial Z} = 0, \tag{5.22}$$

$$\frac{\partial \hat{u}}{\partial T} + \hat{u}\frac{\partial \hat{u}}{\partial X} + \hat{w}\frac{\partial \hat{u}}{\partial Z} = -\frac{\partial \hat{p}}{\partial X} \tag{5.23}$$

and

$$\frac{\partial \hat{p}}{\partial Z} + \frac{gh}{U^2} = O\left(\frac{h^2}{\lambda^2}\right). \tag{5.24}$$

From the last equation, we see that, if $h \ll \lambda$, the fluid inertia terms in the Z-direction can be neglected and also that the appropriate choice for U is $\sqrt{gh}$. Hence, the pressure is *hydrostatic* to lowest order. Thus, reverting to dimensional variables and integrating, we have

$$p = -\rho g z + \rho g \eta + p_0, \tag{5.25}$$

where $z = \eta(x, t)$ is the equation of the surface of the water. Hence, substituting for $\hat{p}$ in (5.23) and writing that equation in dimensional variables, we obtain

$$\frac{du}{dt} = -g\frac{\partial \eta}{\partial x}, \tag{5.26}$$

showing that the convective derivative of u is independent of z. Thus, if u is initially independent of z, it will remain independent of z for all time.[3] Then, writing $u = u(x, t)$, (5.26) reduces to

$$\frac{\partial u}{\partial t} + u\frac{\partial u}{\partial x} + g\frac{\partial \eta}{\partial x} = 0. \tag{5.27}$$

Now we can also integrate (5.22) with respect to z and this leads to

$$w = -\frac{\partial u}{\partial x}z, \tag{5.28}$$

if we assume a flat bottom with $w = 0$ on $z = 0$. Now we finally need to apply the kinematic boundary condition (3.10) at the surface $z = \eta$ to get

$$w = \frac{\partial \eta}{\partial t} + u\frac{\partial \eta}{\partial x}, \tag{5.29}$$

and (5.28) and (5.29) lead to

$$\frac{\partial \eta}{\partial t} + \frac{\partial}{\partial x}(u\eta) = 0. \tag{5.30}$$

Equations (5.27) and (5.30) can be thought of as statements of conservation of momentum and mass, respectively, averaged through the depth of the water; they are two nonlinear equations for $u(x, t)$ and $\eta(x, t)$. It is convenient to write $s^2 = g\eta$ and then add and subtract these two equations to obtain

$$\left(\frac{\partial}{\partial t} + (u \pm s)\frac{\partial}{\partial x}\right)(u \pm 2s) = 0. \tag{5.31}$$

We have now arrived at a formulation of the problem which is very like that obtained in Section 5.2.1 and shows that

$$u \pm 2s = \text{constant}$$

on the characteristics

$$\frac{dx}{dt} = u \pm s.$$

In fact, the model (5.31) is *identical* with the one-dimensional unsteady gas dynamic equations (5.9) if we put $c = s$ and $\gamma = 2$ (remember $\gamma = 1.4$ for air).

Reviewing the three sets of equations (5.7), (5.8) and (5.11)–(5.13) and (5.27), (5.30), we can look at a more general methodology for dealing with these systems. Each of these sets of equations can be written in the form

$$A\frac{\partial \mathbf{u}}{\partial X} + B\frac{\partial \mathbf{u}}{\partial Y} = \mathbf{0}, \tag{5.32}$$

---

[3] It is interesting to relate this assumption to that of irrotationality (Exercise 5.8).

where $\mathbf{u}$ is an n-vector and A and B are $n \times n$ matrices whose entries are functions of the components of $\mathbf{u}(X, Y)$. We note that if A is non-singular and $A^{-1}B$ has an eigenvalue λ and a left eigenvector $\mathbf{l}$, so that

$$\det(\lambda A - B) = 0 \text{ and } \mathbf{l}.A^{-1}B = \lambda \mathbf{l},$$

then

$$\mathbf{l}.\frac{\partial \mathbf{u}}{\partial X} + \lambda \mathbf{l}.\frac{\partial \mathbf{u}}{\partial Y} = 0.$$

Thus, $\int \mathbf{l}.d\mathbf{u}$ is constant on the curve $dY/dX = \lambda$, and this is exactly equivalent to the procedure that we have used already for each of the above problems. Indeed, this is the starting point for the theory of *hyperbolic systems*; (5.32) is called hyperbolic if all n eigenvalues of $A^{-1}B$ are real and distinct, $dY/dX = \lambda$ are the characteristic curves and $\int \mathbf{l}.d\mathbf{u}$ are the Riemann invariants [43].

Finally, we remark that if we linearise the shallow water equations (5.27) and (5.30) by assuming that u and $\eta - h$ are small[4], we are led to the familiar one-dimensional wave equation

$$\frac{\partial^2 \eta}{\partial t^2} = gh\frac{\partial^2 \eta}{\partial x^2}. \tag{5.33}$$

Solutions of (5.33) are called *tidal waves*, and it is easy to verify that in most seas and oceans, the tides move along the shore at a speed of $O(\sqrt{gh})$. It has already been observed in Section 4.4.1 that, as $h \to 0$ in the Stokes wave solution (4.27), the dispersion relation reduces to

$$\omega^2 = ghk^2,$$

exactly as predicted from (5.33). Thus, long, small-amplitude waves on shallow water are non-dispersive and the phase and group velocities for such waves are both $\sqrt{gh}$. In fact, most tidal waves propagate in two dimensions and in this case (5.33) is replaced by

$$\frac{\partial^2 \eta}{\partial t^2} = gh\left(\frac{\partial^2 \eta}{\partial x^2} + \frac{\partial^2 \eta}{\partial y^2}\right),$$

which is also the equation for waves on a membrane.

---

[4] We can be more precise about how small these quantities are; if we define $(\eta - h)/h = O(\varepsilon)$, then u will be $O(\varepsilon\sqrt{gh})$.

5.3 *Nonlinearity and Dispersion

5.3.1 The Korteweg–de Vries Equation

The full model for nonlinear surface waves, described by (3.8), (3.10) and (3.11), can be simplified in other parameter regimes that are distinct from Stokes waves, shallow water theory or tidal theory, and a fascinating situation occurs if we consider the *long time evolution* of a tidal wave. To illustrate why the solution obtained in the last section may not be valid over long times, we first consider the model equation

$$\frac{\partial \phi}{\partial t} + \frac{\partial \phi}{\partial x} = \varepsilon \frac{\partial^2 \phi}{\partial x^2}, \tag{5.34}$$

where ε is a small parameter. If we write $\phi \sim \phi_0 + \varepsilon \phi_1 + \cdots$ and expand in powers of ε, we find that

$$\phi_0 = f(x - t)$$

and

$$\frac{\partial \phi_1}{\partial t} + \frac{\partial \phi_1}{\partial x} = f''(x - t),$$

so that

$$\phi_1 = t f''(x - t) + g(x - t),$$

where f and g are arbitrary functions. Thus, we see that, when $t \sim O(\varepsilon^{-1})$, $\varepsilon \phi_1$ will be of the same order as ϕ_0 and the expansion for small ε is no longer valid. To find a solution valid for such timescales, we need to rewrite (5.34) in terms of new independent variables $\xi = x - t$ and $\tau = \varepsilon t$ so that it becomes

$$\frac{\partial \phi}{\partial \tau} = \frac{\partial^2 \phi}{\partial \xi^2}. \tag{5.35}$$

Now we need to solve (5.35) with the initial condition $\phi = f(\xi)$ at $\tau = 0$ to get a solution to (5.34) which is valid for all times. It is interesting to note that, for this linear model, we can use Fourier transforms to obtain the long-time behaviour and thereby assess the validity of (5.35) (Exercise 5.7).

We now show how the same method can be applied to tidal waves. We recall that Stokes waves were derived under the assumption that the ratio of amplitude to depth, $a/h = \varepsilon$, is small. Shallow water theory assumes that the ratio of depth to wavelength, $h/\lambda = \delta$, is small, and tidal wave theory assumes that both ε and δ are small. Using the same scalings as tidal wave theory in Section 5.2.3, we nondimensionalise x with λ, z with h, η with a, t with $\lambda/\sqrt{gh}$ and u with $\varepsilon\sqrt{gh}$. Then, the appropriate scaling for ϕ is $\varepsilon\lambda\sqrt{gh}$ and (3.8), (3.10) and (3.11) become, in nondimensional form,

$$\frac{\partial^2 \phi}{\partial z^2} + \delta^2 \frac{\partial^2 \phi}{\partial x^2} = 0, \tag{5.36}$$

with

$$\frac{\partial \phi}{\partial t} + \eta + \frac{\varepsilon}{2\delta^2} \left(\left(\frac{\partial \phi}{\partial z} \right)^2 + \delta^2 \left(\frac{\partial \phi}{\partial x} \right)^2 \right) = 0 \qquad (5.37)$$

and

$$\frac{\partial \phi}{\partial z} = \delta^2 \frac{\partial \eta}{\partial t} + \varepsilon \delta^2 \frac{\partial \phi}{\partial x} \cdot \frac{\partial \eta}{\partial x} \qquad (5.38)$$

on $z = \varepsilon\eta$. The final boundary condition on the bottom is

$$\frac{\partial \phi}{\partial z} = 0 \qquad (5.39)$$

on $z = -1$. We can see at once that we can retrieve Stokes waves by taking $\delta = 1$ and $\varepsilon = 0$.

When both ε and δ are small, we find that, in order to satisfy (5.36) and (5.39), we must write

$$\phi(x, z, t; \varepsilon, \delta) \sim \phi_0(x, t; \varepsilon) + \delta^2 (A(x, t; \varepsilon) - (\frac{1}{2}z^2 + z)\frac{\partial^2 \phi_0}{\partial x^2}) + \cdots, \qquad (5.40)$$

for some function A. Then, writing $\eta \sim \eta_0(x, t; \varepsilon) + \delta^2 \eta_1(x, t; \varepsilon) + \cdots$, equations (5.37) and (5.38) lead to

$$\frac{\partial \phi_0}{\partial t} + \eta_0 = O(\varepsilon, \delta^2)$$

and

$$\frac{\partial^2 \phi_0}{\partial x^2} + \frac{\partial \eta_0}{\partial t} = O(\varepsilon, \delta^2),$$

and hence, when $\varepsilon \ll 1$, to the tidal wave equation (5.33). When $\varepsilon = 1$ and $\delta \to 0$, we can derive the equations for shallow water, (5.26) and (5.30), in a similar way (Exercise 5.8). However, the example (5.34) leads us to examine whether the tidal wave approximation will really be valid for large times. We see that as t increases, the time derivative in (5.37) will become as small as some of the neglected terms and, to deal with this, we have to take into account even more terms in the expansion (5.40) for ϕ.

Again using (5.36) and (5.39), but going to the next term in the expansion in δ^2, gives

$$\phi \sim \phi_0(x, t; \varepsilon) + \delta^2 \left(A(x, t; \varepsilon) - \left(\frac{1}{2}z^2 + z \right) \frac{\partial^2 \phi_0}{\partial x^2} \right)$$

$$+ \delta^4 \left(B(x, t; \varepsilon) - \left(\frac{1}{2}z^2 + z \right) \frac{\partial^2 A}{\partial x^2} + (\frac{1}{24}z^4 + \frac{1}{6}z^3 - \frac{1}{3}z) \frac{\partial^4 \phi_0}{\partial x^4} \right) + \cdots.$$

$$(5.41)$$

Then, remembering that the surface condition is applied on $z = \varepsilon\eta$, (5.37) is

$$\frac{\partial\phi_0}{\partial t} + \eta_0 + \frac{1}{2}\varepsilon\left(\frac{\partial\phi_0}{\partial x}\right)^2 + \delta^2\left(\frac{\partial A}{\partial t} + \eta_1\right) = O(\varepsilon\delta^2) \qquad (5.42)$$

and (5.38) gives

$$\frac{\partial\eta_0}{\partial t} + \frac{\partial^2\phi_0}{\partial x^2} + \varepsilon\left(\eta_0\frac{\partial^2\phi_0}{\partial x^2} + \frac{\partial\phi_0}{\partial x}\frac{\partial\eta_0}{\partial x}\right)$$
$$+ \delta^2\left(\frac{\partial^2 A}{\partial x^2} + \frac{1}{3}\frac{\partial^4\phi_0}{\partial x^4} + \frac{\partial\eta_1}{\partial t}\right) + O(\varepsilon\delta^2,\delta^4) = 0. \qquad (5.43)$$

At this point, we need to rescale $t = \varepsilon^{-1}\tau$ and, remembering that the tidal wave generates solutions for η_0 and ϕ_0 of the form $f(x-t) + g(x+t)$, we follow just the right-travelling wave by writing $x - t = \xi$ and keeping ξ of $O(1)$. We also have to decide on the relation between our two small parameters ε and δ. It is clear from (5.42) and (5.43) that the most interesting case will be when $\varepsilon = O(\delta^2)$ and accordingly we write $\delta^2 = \kappa\varepsilon$. We thus write (5.42), (5.43) as

$$-\frac{\partial\phi_0}{\partial\xi} + \eta_0 + \varepsilon\left(\frac{1}{2}\left(\frac{\partial\phi_0}{\partial\xi}\right)^2 - \kappa\frac{\partial A}{\partial\xi} + \kappa\eta_1 + \frac{\partial\phi_0}{\partial\tau}\right) = O(\varepsilon^2)$$

and

$$-\frac{\partial\eta_0}{\partial\xi} + \frac{\partial^2\phi_0}{\partial\xi^2} + \varepsilon\left(\eta_0\frac{\partial^2\phi_0}{\partial\xi^2} + \frac{\partial\phi_0}{\partial\xi}\frac{\partial\eta_0}{\partial\xi}\right.$$
$$\left. +\kappa\frac{\partial^2 A}{\partial\xi^2} + \frac{\kappa}{3}\frac{\partial^4\phi_0}{\partial\xi^4} - \kappa\frac{\partial\eta_1}{\partial\xi} + \frac{\partial\eta_0}{\partial\tau}\right) = O(\varepsilon^2).$$

Hence, $\eta_0 = \partial\phi_0/\partial\xi$, and, by adding the ξ-derivative of the first equation to the second, we finally obtain that

$$\frac{\partial\eta_0}{\partial\tau} + \frac{3}{2}\eta_0\frac{\partial\eta_0}{\partial\xi} + \frac{\kappa}{6}\frac{\partial^3\eta_0}{\partial\xi^3} = 0 \qquad (5.44)$$

is the condition for (5.42), (5.43) to be simultaneously valid.[5] This is the *Korteweg–de Vries (KdV) equation* and it is valid for times of $O((\lambda^3/h^3)\sqrt{h/g})$ if we assume that $a/h = O((h/\lambda)^2)$. The equation represents a balance between the linear "tidal wave" term $\partial\eta_0/\partial\tau$, the nonlinear "shallow water" term $\eta_0(\partial\eta_0/\partial\xi)$ and the dispersive "Stokes wave" term $\partial^3\eta_0/\partial\xi^3$.

The KdV equation has travelling wave solutions which can be obtained by writing $\eta_0 = f(\xi - c\tau)$. Substituting into (5.44) and integrating once gives

---

[5] The step leading to (5.44) is yet another example of the Fredholm Alternative.

$$\frac{\kappa}{6}\frac{d^2f}{d\chi^2} = cf - \frac{3}{4}f^2 + k_1,$$

(5.45)

where k_1 is a constant and $\chi = \xi - c\tau = x - (1 + c\varepsilon)t$. If we require that, as $\chi \to \pm\infty$, the solution f and its derivatives tend to zero, then $k_1 = 0$ and (5.45) can be integrated to give

$$f = 2c\,\mathrm{sech}^2\left(\sqrt{\frac{3c}{2\kappa}}(\chi + d)\right),$$

(5.46)

where d is a constant. Because of its behaviour as $|\chi| \to \infty$, this profile is called a *solitary wave*. It is a travelling wave of constant speed and shape which can easily be observed in either a long straight canal[6] or in a laboratory. Even more remarkable is the *theory of solitons* which was opened up by the study of the KdV equation. A soliton is a solitary wave that has certain very special properties in addition to its permanent shape, and (5.46) was the first solitary wave that was discovered to have these special properties. We will return to this topic briefly at the end of this section after we have considered another important example of gradual nonlinear modulation.

We remark that, although nonlinearity destroys the predictions of linear tidal theory for long times, it does not destroy most of the predictions of linear Stokes wave theory on deeper water. Nonetheless, interesting new phenomena occur when we consider small nonlinear corrections to Stokes waves, for example, by setting $\delta = 1$ in (5.36)–(5.38) and expanding for small ε. When this is done for a propagating wave train that is harmonic in time, the quadratic terms in the free boundary conditions have a non-zero time average and so inevitably induce a *steady* correction to the velocity field. This is an example of *acoustic streaming* (see Riley [48]), which also predicts a small steady force on an obstacle placed beneath a harmonic wave train.

However, an even more dramatic fate awaits the harmonic wave trains of Stokes waves over long times as we now show.

5.3.2 The Nonlinear Schrödinger Equation

Another situation in which the linear approximation breaks down is when we consider the periodic solution for Stokes waves on deep water over a long time. Once again we illustrate the way this happens by considering a simple model equation — in this case, an ordinary differential equation with a periodic solution in the linear approximation and a seemingly small quadratic nonlinearity. Thus, we consider the equation

---

[6] Scott Russell famously observed "a great wave of elevation" while riding his horse along the towpath of a canal in 1845 ([49]).

$$\frac{d^2x}{dt^2} + x = \varepsilon x^2,$$ (5.47)

and put $x \sim x_0 + \varepsilon x_1 + \varepsilon^2 x_2 + \cdots$ to obtain the equations

$$\frac{d^2x_0}{dt^2} + x_0 = 0,$$ (5.48)

$$\frac{d^2x_1}{dt^2} + x_1 = x_0^2$$ (5.49)

and

$$\frac{d^2x_2}{dt^2} + x_2 = 2x_0x_1.$$ (5.50)

Solving (5.48) and (5.49) gives

$$x_0 = Ae^{it} + A^*e^{-it}$$ (5.51)

and

$$x_1 = Be^{it} + B^*e^{-it} - \frac{1}{3}A^2e^{2it} + 2AA^* - \frac{1}{3}A^{*2}e^{-2it},$$ (5.52)

where A and B are complex constants and * denotes complex conjugate.[7] Now we find that equation (5.50) becomes

$$\frac{d^2x_2}{dt^2} + x_2 = \frac{10}{3}A^2A^*e^{it} + \frac{10}{3}AA^{*2}e^{-it} + \text{ constant } + \text{ terms in } e^{\pm 2it}, e^{\pm 3it}.$$

Hence, there will be terms of the form $te^{\pm it}$ in x_2 and these will grow with time so that our expansion for x will become invalid when $\varepsilon^2 t = O(1)$. Since the solution will inevitably contain oscillations on a timescale of $O(1)$, the procedure here is to introduce a slow time variable $T = \varepsilon^2 t$ and regard x as a function of *both* t and T. This method of multiple scales was used for spatial variables in Section 4.6.2 and in practice; all we need to do to use the method is to again use the chain rule to replace d/dt by $\partial/\partial t + \varepsilon^2 \partial/\partial T$ so that (5.47) becomes

$$\frac{\partial^2 x}{\partial t^2} + 2\varepsilon^2 \frac{\partial^2 x}{\partial t \partial T} + \varepsilon^4 \frac{\partial^2 x}{\partial T^2} + x = \varepsilon x^2.$$

Now, expanding in powers of ε again, x_0 and x_1 take exactly the same forms (5.51) and (5.52) as before as long as we now regard A and B as functions of T. The equation for x_2 becomes

$$\frac{\partial^2 x_2}{\partial t^2} + x_2 = 2x_0x_1 - \frac{2\partial^2 x_0}{\partial t \partial T},$$

---

[7] We now use notation (5.51) rather than $x_0 = \text{Rl}(Ae^{it})$ in order to avoid confusion in evaluating the nonlinear terms when it is important to note that $\text{Rl}(A^2) \neq (\text{Rl}\,A)^2$.

and we can eliminate the offending terms in $e^{\pm it}$ on the right-hand side if we make A satisfy the equation

$$2i\frac{dA}{dT} = \frac{10}{3}A^2A^*. \tag{5.53}$$

Solutions of this equation for A will give x_0 as a slowly modulated oscillatory function which is a valid asymptotic approximation for x for times of $O(\varepsilon^{-2})$ (Exercise 5.10).

Now let us apply these ideas to the periodic Stokes wave trains that we considered in Chapter 3. We can obtain the nondimensional form of the equations by putting $\delta = 1$ in equations (5.36)–(5.38), and for simplicity, we consider the case of infinitely deep water.

Schematically, we can write this system as

$$\frac{\partial^2 \phi}{\partial x^2} + \frac{\partial^2 \phi}{\partial z^2} = 0, \tag{5.54}$$

with

$$\frac{\partial^2 \phi}{\partial t^2} + \frac{\partial \phi}{\partial z} = \varepsilon Q \tag{5.55}$$

on $z = 0$, where Q is a power series in ε involving terms that are nonlinear in ϕ and η. The precise form of Q will be needed if the subsequent analysis is to be followed in detail, and a recipe for it is given in Exercise 5.12. However, readers who simply want the general gist do not need this information if they are prepared to trust the authors' calculations!

Taking the wave number k to be positive without loss of generality, the linear Stokes wave train for deep water can be written as

$$\phi_0 = (Ae^{i(kx-\omega t)} + A^*e^{-i(kx-\omega t)})e^{kz}, \tag{5.56}$$

where $\omega^2 = k$ and, exactly as in model problem above, we see that when we put $\phi \sim \phi_0 + \varepsilon\phi_1 + \varepsilon^2\phi_2 + \cdots$, ϕ_2 will grow algebraically in time. Thus, the expansion is invalid when $\varepsilon^2 t = O(1)$. Because we are now dealing with functions of several variables, it turns out that we need to introduce the four slow variables $X_1 = \varepsilon x$, $X_2 = \varepsilon^2 x$, $T_1 = \varepsilon t$ and $T_2 = \varepsilon^2 t$, but otherwise we proceed as before. If we regard ϕ as a function of $x, z, t, X_1, T_1, X_2, T_2$, the equation for ϕ_0 is unchanged except that A will now be a function of X_1, T_1, X_2, T_2. However, the equation for ϕ_1 becomes

$$\frac{\partial^2 \phi_1}{\partial x^2} + \frac{\partial^2 \phi_1}{\partial z^2} = -2\frac{\partial^2 \phi_0}{\partial x \partial X_1}, \tag{5.57}$$

with

$$\frac{\partial^2 \phi_1}{\partial t^2} + \frac{\partial \phi_1}{\partial z} = -2\frac{\partial^2 \phi_0}{\partial t \partial T_1} - 2\frac{\partial \phi_0}{\partial x}\frac{\partial^2 \phi_0}{\partial x \partial t} - 2\frac{\partial \phi_0}{\partial z}\frac{\partial^2 \phi_0}{\partial z \partial t} \tag{5.58}$$

on $z = 0$, and so

$$
\phi_1 = \left(-iz\frac{\partial A}{\partial X_1}e^{i(kx-\omega t)} + iz\frac{\partial A^*}{\partial X_1}e^{-i(kx-\omega t)} \right) e^{kz}
$$
$$
+ \left(Be^{i(kx-\omega t)} + B^*e^{-i(kx-\omega t)} \right) e^{kz},
$$

where B depends on X_1, T_1, X_2, T_2. Using the boundary condition (5.58), we find that the elimination of terms in $e^{\pm i(kx-\omega t)}$ requires[8]

$$
\frac{\partial A}{\partial X_1} + 2\omega\frac{\partial A}{\partial T_1} = 0.
$$

Thus, A is a function of $X_1 - VT_1, X_2$ and T_2, where $V = 1/(2\omega)$ and, remembering that $\omega^2 = k$, we can identify V as the group velocity $d\omega/dk$. This observation points to a very general result in the theory of modulated linear wave trains that says that we can describe the gradual effect of nonlinearity if and only if we work in a frame moving with the group velocity.

Finally, we set $\xi = X_1 - VT_1$ and carry on to the next term in the expansion for ϕ. We find that

$$
\phi_2 = \left(-\frac{1}{2}\frac{\partial^2 A}{\partial\xi^2}z^2 - i\left(\frac{\partial A}{\partial X_2} + \frac{\partial B}{\partial\xi} \right)z + C \right) e^{i(kx-\omega t)+kz} + \text{complex conjugate},
$$

where C is a function of ξ, X_2, T_2. When we use the boundary condition (5.55) to $O(\varepsilon^2)$, we find that the terms in $e^{\pm i(kx-\omega t)}$ will balance only if A satisfies the *Nonlinear Schrödinger Equation* (NLS equation)

$$
i\left(\frac{\partial A}{\partial T_2} + V\frac{\partial A}{\partial X_2} \right) - V^3\frac{\partial^2 A}{\partial\xi^2} + cA^2A^* = 0, \tag{5.59}
$$

where c is a real constant (see Exercise 5.12). This equation can be simplified by transforming to a frame moving with the group velocity on the T_2-scale and writing $\zeta = X_2 - VT_2, \tau = T_2$, so that as a function of ξ, ζ and τ, A satisfies

$$
i\frac{\partial A}{\partial\tau} - V^3\frac{\partial^2 A}{\partial\xi^2} + cA^2A^* = 0, \tag{5.60}
$$

which is the more familiar form of the NLS equation.

Despite its apparent complexity, this equation has been much studied and as much is known about it as is known about the KdV equation. Notice the following:

(i) If the dispersive term $\partial^2 A/\partial\xi^2$ is absent, the NLS equation effectively reduces to (5.53).
(ii) If the nonlinear terms are neglected, the NLS equation reduces to what is effectively the beam equation (Exercise 4.12) or the parabolic wave equation (4.86).
(iii) There is a particular solution of the form

---

[8] Note that if we had introduced the variable $T_1 = \varepsilon t$ in the solution of our model equation (5.47), we would merely have found that $\partial A/\partial T_1 = 0$.

$$A = \left(-\frac{2V^3}{c}\right)^{1/2} e^{i\beta\xi - iV^3\tau} \operatorname{sech}\xi, \qquad (5.61)$$

where β is any real constant. This solution can be written in terms of (x, t) as

$$A = \left(-2\frac{V^3}{c}\right)^{1/2} e^{i\varepsilon^2(\beta x - (\beta V + V^3)t)} \operatorname{sech}\varepsilon(x - Vt),$$

which illustrates the modulation in phase and amplitude that governs the eventual fate of the Stokes wave train (5.56).

(iv) We can use (5.60) to assess the stability of trains of Stokes waves. A solution that represents such a wave train is

$$A = A_s = \sqrt{\frac{-\Omega}{c}} e^{-i\Omega\tau},$$

and, when we write $A = A_s + \varepsilon A_1 + \cdots$ in (5.60) and neglect higher order terms (Exercise 5.14), we find that A_1 is proportional to $\exp(-i\Omega\tau \pm i(\kappa\xi - \sigma\tau))$, where

$$\sigma^2 = V^3\kappa^2(V^3\kappa^2 - 2\Omega).$$

Thus, σ has a non-zero imaginary part for sufficiently small values of the wave number κ, and this leads to the so-called *Benjamin–Feir instability*. As explained below, (5.61) describes part of the eventual evolution of these unstable waves.

Although (5.61) can be likened to the solitary wave solution (5.46) of the KdV equation, we must remember that A is now the modulation of the wave train (5.56), so we have a so-called envelope solitary wave. However, there is a much more profound mathematical unity between (5.44) and (5.60) because both equations are susceptible to the theory of *inverse scattering* (Drazin and Johnson [17]). In fact, we have uncovered just the tip of an intellectual iceberg, and (5.46) and (5.61) have far more interesting attributes than simply being spatially localised solitary waves. Hence, they have earned the sobriquet of *solitons*, about which far more can be found in specialised texts such as Dodd et al. [16] which study partial differential equations that are *completely integrable* in the Hamiltonian sense. Both the KdV equation and the NLS equation are completely integrable and their solution can, in principle, be written down when appropriate data is prescribed at $\tau = 0$ for all ξ. A remarkable feature is that most reasonable initial data will lead to a universal behaviour as $\tau \to \infty$ in which the solution will consist of a series of solitary waves of the form (5.46) or (5.61).[9] Their more specialised name of solitons reflects the fact that, for large times, they move like independent particles, even to the extent of "passing through" each other while retaining their own amplitude and velocity after each

---

[9] Note that we are, at this stage of the book, unable to make any comparable statement about the relatively trivial (5.1).

"collision". The theory of inverse scattering can predict the size and number of these solitons in any particular situation.

5.3.3 Resonance

In the previous sections, we have discussed water waves that can propagate in unbounded space and time. When water is contained in an oscillating tank, the problem is complicated by multiple reflections from the boundaries, although the linear time-periodic theory described in Chapter 3 will still hold unless the frequency of the forcing oscillations is close to a natural frequency of the system. If this is the case, resonance occurs as described in Section 4.2 and Exercise 3.2, and the amplitude of the waves predicted by linear theory becomes unbounded. However, when the difference between the frequency of the forcing oscillation and a natural frequency of the system (this is called the detuning parameter) is small, the periodic response may be bounded, even for zero detuning, as long as dispersion or nonlinearity is taken into account.

To understand the effect of nonlinearity on the resonant response it is easiest to begin by looking at the generalisation of (5.47) in the form of a forced oscillator

$$\frac{d^2x}{dt^2} + (1 + \delta)x - \varepsilon x^2 = \cos t, \tag{5.62}$$

where the detuning δ is small so that the forcing frequency is close to the natural frequency$(1 + \delta)^{1/2}$. When the nonlinearity is cubic rather than quadratic, (5.62) is called *Duffing's Equation*. Note that if $\varepsilon = 0$, the periodic response has amplitude $1/|\delta|$.

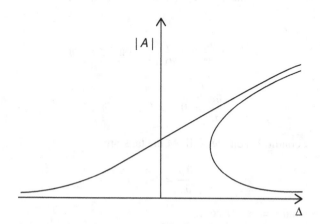

Fig. 5.2 Response diagram for solutions of (5.64) when $\varepsilon > 0$.

It turns out that, as in the analysis of (5.47), we need to prevent growth in the third term in the expansion in powers of ε. For this to be possible, the nonlinearity must be comparable to the detuning effects at the third order and this can only happen when $\delta = \Delta\varepsilon^{2/3}$, where $\Delta = O(1)$. Then, we can write $x \sim \varepsilon^{-2/3}x_0 + \varepsilon^{-1/3}x_1 + x_2 + \cdots$, where x_0 and x_1 satisfy equations (5.48) and (5.49). However, x_2 now satisfies

$$\frac{d^2 x_2}{dt^2} + x_2 = -\Delta x_0 - 2x_0 x_1 + \cos t. \tag{5.63}$$

Using the solutions (5.51) and (5.52), we find the condition needed to avoid secular terms in (5.63) is

$$\frac{10}{3}A^3 - \Delta A + \frac{1}{2} = 0. \tag{5.64}$$

The resulting *response curve* of $|A|$ as a function of the detuning Δ is sketched in Figure 5.2; note that there is a phase shift of π between the two branches and that the curves bend to the left rather than the right if the sign of the nonlinear term in (5.62) is reversed.

It is now, in principle, easy to study the sloshing of water in a rectangular tank which is subject to small amplitude horizontal oscillations of frequency ω. We take the depth of the water ah to be comparable in size to the width of the tank πa and assume the amplitude of the forcing oscillations is $a\varepsilon$ where $\varepsilon \ll 1$. Starting from (3.8), (3.10), (3.11) and (3.14), we nondimensionalise x and z with a, ϕ with $\varepsilon a^2 \omega$, η with εa and t with $1/\omega$ to yield

$$\frac{\partial^2 \phi}{\partial x^2} + \frac{\partial^2 \phi}{\partial z^2} = 0, \tag{5.65}$$

with

$$\frac{\partial \phi}{\partial t} + \frac{g}{a\omega^2}\eta + \frac{1}{2}\varepsilon\left(\left(\frac{\partial \phi}{\partial x}\right)^2 + \left(\frac{\partial \phi}{\partial z}\right)^2\right) = 0 \tag{5.66}$$

and

$$\frac{\partial \phi}{\partial z} = \frac{\partial \eta}{\partial t} + \varepsilon\frac{\partial \phi}{\partial x}\frac{\partial \eta}{\partial x} \quad \text{on} \quad z = \varepsilon\eta \tag{5.67}$$

and

$$\frac{\partial \phi}{\partial z} = 0 \quad \text{on} \quad z = -h. \tag{5.68}$$

The boundary conditions on the walls of the tank are

$$\frac{\partial \phi}{\partial x} = -\sin t \tag{5.69}$$

on $x = \varepsilon \cos t$ and $x = \pi + \varepsilon \cos t$.

When there is no forcing and $\varepsilon = 0$, the nth normal mode of oscillation is given by

$$\phi_n = \cos nx \cosh n(z + h)(a_n \cos \lambda_n t + b_n \sin \lambda_n t),$$

where a_n, b_n are constant and the natural frequency λ_n is given by

$$\lambda_n^2 = \frac{ng}{a\omega^2}\tanh nh.$$

Resonance will occur when the forcing frequency is close to one of the λ_n, and we will consider just the first harmonic and write $\lambda_1^2 = 1+\delta$ or $\omega^2=g(\tanh h)/a(1+\delta)$. Now we find that, when the detuning parameter δ is small, the approximate solution of the linear problem is, as shown in Exercise 5.13,

$$\phi = \frac{4}{\delta\pi\cosh h}\cos x \cosh(z+h)\sin t + O(1),$$

where the $O(1)$ term satisfies the boundary condition (5.69).

By analogy with (5.62), we find that we can analyse the solution when δ is as large as $O(\varepsilon^{2/3})$ by proceeding in exactly the same way. We take $\delta = \varepsilon^{2/3}\Delta$ and write

$$\phi \sim \varepsilon^{-2/3}\phi_0 + \varepsilon^{-1/3}\phi_1 + \phi_2 + \cdots$$

and

$$\eta \sim \varepsilon^{-2/3}\eta_0 + \varepsilon^{-1/3}\eta_1 + \eta_2 + \cdots,$$

where

$$\phi_0 = A\cos x\cosh(z+h)\sin t$$

and

$$\eta_0 = -A\cos x\sinh h\cos t,$$

where A is to be determined. After a long calculation [44], we find that the condition for there to be no secular terms in ϕ_2 is

$$H(h)A^3 + \Delta\sinh h\, A - \frac{4}{\pi}\tanh h = 0, \tag{5.70}$$

where

$$H = -\frac{1}{32}(9 + 15\mathrm{sech}^2 h - 8\sinh^6 h)\mathrm{sech}^2 h\,\mathrm{cosech}\, h.$$

Thus, the amplitude A satisfies an equation similar to (5.64) and its dependence on Δ takes the same form as in Figure 5.2 when $H < 0$. However, $H(h)$ becomes positive as h increases through $h = h_* \simeq 1.06$ and so, for deeper water, the response curve bends to the left. Moreover, when $h \simeq h_*$, an even lengthier calculation shows that A satisfies a quintic equation.

We emphasise that the above result no longer applies when $h \to 0$ and we approach the shallow water limit. It can be shown that as h decreases, the nonlinearity induces more and more branches in the response diagram Figure 5.2, eventually culminating in a band of solutions with jump discontinuities or shocks.[10] We recall

---

[10] As mentioned before, solutions containing discontinuities or shocks will be considered in detail in Chapter 6.

that in Section 5.2.3, we saw that the limit $h \to 0$ reduced the model to that of a one-dimensional organ pipe with $\gamma = 2$ and the limiting case illustrated in Figure 5.3 applies equally to resonant waves in an organ pipe [45].

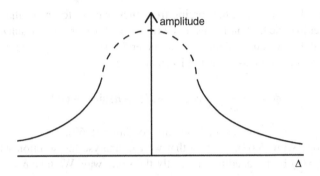

Fig. 5.3 Response diagram for solutions of (5.70) as $h \to 0$; the dotted curve indicates solutions containing shocks.

Finally, we remark that water in a tank can also be forced to resonate by applying small *vertical* oscillations at a resonant frequency. The analysis of this situation is too complicated to describe here, but a crude model is to simply consider gravity as an oscillatory function of time. This enables a small-amplitude asymptotic analysis to be carried out along the lines adopted for the Mathieu equation (4.49) in Exercise 4.13. This reveals that the resonance has a structure that is quite different from traditional resonances in which the response of a system with discrete normal modes becomes unexpectedly large when the forcing frequency is close to a natural frequency (as was the case of "horizontal sloshing" described above). For vertically forced sloshing, it turns out that the amplitude of each normal mode does indeed satisfy a Mathieu equation and hence, from Figure 4.5, unexpectedly large responses can occur over *bands* of forcing frequencies and this situation is often called *parametric resonance* [4]. The solutions are only precisely periodic at the boundaries of the unshaded "Arnold tongues" in Figure 4.5, and it can be shown that these periodic solutions play a key role when weak nonlinearity is taken into account.

5.4 Smooth Solutions for Hyperbolic Models

We now present prototypical solutions of the three hyperbolic models described in Sections 5.2.1 to 5.2.3, each of which relies on the existence of a region of *simple wave flow*. This concept describes any situation in which the region of interest is adjacent to a *uniform region* in which all the flow variables are constant, and this enables us to exploit our knowledge of the Riemann invariants. The region of interest will be the simple wave region that is spanned by one family of characteristics that emanate from the uniform region. Thus, the corresponding Riemann invariant will take a known constant value *everywhere* in the region of interest. Hence, the problem is immediately reduced to a scalar first-order partial differential equation which can be solved by traditional methods. To make things even easier, we can note that the constancy of the second Riemann invariant along characteristics of the second family implies that these characteristics will be straight lines and that the flow variables will be constant along these lines. We now use this strategy to obtain exact analytic solutions to three famous problems.

5.4.1 The Piston Problem for One-dimensional Unsteady Gas Dynamics

We suppose that gas is at rest with speed of sound c_0 in $x < 0$ in a tube $-\infty < x < \infty$ containing a piston at $x = 0$. At $t = 0$, the piston begins to move in the positive x-direction with given velocity $\dot{X}(t)$. The flow will be homentropic and (5.9) will therefore hold in $x < X(t)$, where $x = X(t)$ is the position of the piston at time t. We first assume that the gas is being expanded so that $\dot{X}(t)$ is a monotonically increasing function of t and $X(0) = \dot{X}(0) = 0$.

In the region $x < 0$, $t < 0$, $u = 0$ and $c = c_0$, so that the characteristics there will be straight lines with slope $\pm c_0$. For $t > 0$, we first consider a point $P(\bar{x}, \bar{t})$, as shown in Figure 5.4, which is at the intersection of two characteristics PA and PB emanating from the region $x < 0, t < 0$. Along PA, $u + 2c/(\gamma - 1) = 2c_0/(\gamma - 1)$ and along PB, $u - 2c/(\gamma - 1) = -2c_0/(\gamma - 1)$, so we can deduce straightaway that $u = 0$ and $c = c_0$ at P and that the characteristics PA and PB are straight lines with slope $\pm c_0$. Hence, P lies in a uniform region in which $u = 0$, $c = c_0$ as long as $x < -c_0 t$; this is just another way of saying that the disturbance caused by the piston propagates into the gas with speed c_0.

When $x > -c_0 t$, we can use the fact that all the *positive* characteristics, labelled C^+ in Figure 5.4, emanate from the uniform region and so

$$u + \frac{2c}{\gamma - 1} = \frac{2c_0}{\gamma - 1} \tag{5.71}$$

holds everywhere in $x < X(t)$. In particular, since we know that $u = \dot{X}(t)$ on the piston, we can deduce that $c = c_0 - (\gamma - 1)/2\dot{X}(t)$ on the piston. This relation only makes physical sense if $c \geq 0$ and so we will, for the moment, impose the further restriction that

$$\dot{X}(t) \leq \frac{2c_0}{\gamma - 1}.$$

As mentioned above, the region $-c_0 t < x < X(t)$ is a region of simple wave flow and, if we consider the solution at the point Q in this region, we may use the fact that the negative characteristic through Q will be a straight line along which u and c are both constant. We suppose that C is the point where $t = \tau$ and $x = X(\tau)$ so that,

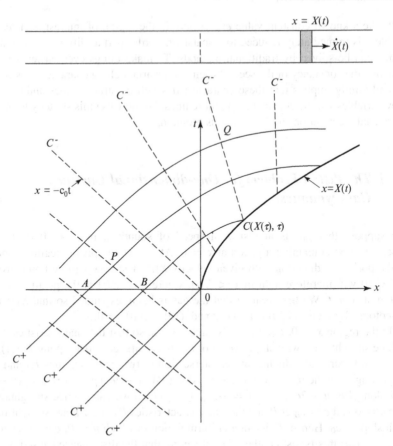

Fig. 5.4 The piston problem: — positive characteristics, - - - negative characteristics.

along QC,

$$u = \dot{X}(\tau) \quad \text{and} \quad c = c_0 - \frac{(\gamma - 1)}{2}\dot{X}(\tau). \tag{5.72}$$

Since QC is a negative characteristic, labelled C^- in Figure 5.4, on which $dx/dt = u - c$, the equation for QC is

$$x - X(\tau) = (t - \tau)\left(\frac{\gamma + 1}{2}\dot{X}(\tau) - c_0\right), \tag{5.73}$$

and thus the solution for u, c is given parametrically by equations (5.72) and (5.73). This solution is valid for all points x, t which satisfy

$$-c_0 t < x < X(t),$$

provided that $0 < \dot{X}(\tau) < 2c_0(\gamma - 1)$. If $\dot{X}(\tau)$ exceeds $2c_0/(\gamma - 1)$, a vacuum will form between the piston and the gas, which will expand freely in such a way that it is bounded by the characteristic with slope $2c_0/(\gamma - 1)$ on which $c = 0$. The solution given by (5.72) and (5.73) will still be valid behind this characteristic.

By considering functions $X(t)$ such that $\ddot{X}(0)$ becomes larger and larger while $\dot{X}(t)$ tends to a constant V for $t > 0$, an analytic solution can also be found to the problem when the piston is started impulsively and moves out of the tube with speed $V \leq 2c_0/(\gamma - 1)$. In this limit, the C^- characteristics more and more nearly fan out from the origin in Figure 5.4 and, in the limit, there is an *expansion fan* or *centred simple wave* between two regions of uniform flow, as shown in Figure 5.5. In the expansion fan, the straight negative characteristics all go through the origin, and the solution in the fan is such that u and c depend only on x/t. This fact can also be deduced directly from a dimensional argument since there is no time scale in this problem, whose solution can therefore only involve the parameters V and c_0 (Exercise 5.17).

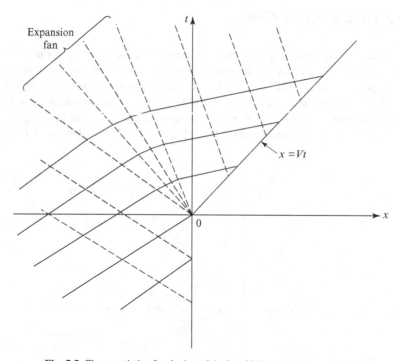

Fig. 5.5 Characteristics for the impulsively withdrawn piston problem.

If the piston moves *into* the fluid, we immediately encounter a completely different scenario. The slope of the negative characteristics QC is such that these characteristics will inevitably intersect. This causes the above solution to break down because u will become multivalued, as described in Section 5.1. We can then only obtain a single-valued solution by introducing a discontinuity into u and c which is known as a shock wave and this will be discussed in Chapter 6. However, if the piston merely accelerates from rest, the smooth flow that exists up until the time the characteristics intersect can still be calculated (see Exercise 5.16).

We conclude with the following observation. Equations (5.9) are of the form $\partial r_i/\partial t + \lambda_i \partial r_i/\partial x = 0$, where r_1, r_2 are the Riemann invariants and λ_1, λ_2 are the slopes of the characteristics. These equations can be transformed by regarding r_1, r_2 as independent variables, and this results in the following *linear* set of equations for x, t:

$$\frac{\partial x}{\partial r_2} = \lambda_1 \frac{\partial t}{\partial r_2} \quad \text{and} \quad \frac{\partial x}{\partial r_1} = \lambda_2 \frac{\partial t}{\partial r_1}.$$

This formulation is of interest in connection with the hodograph transformation which will be described in Section 5.5 and can also be used to solve the problem of intersecting simple waves (Exercise 5.21).

5.4.2 Prandtl–Meyer Flow

The next, slightly less simple, example concerns the flow of a two-dimensional steady supersonic flow past a continuous convex corner. We suppose that a uniform supersonic flow with Mach number M_1 flows parallel to a wall along $y = 0$ in $x < 0$ and that the corner starts smoothly at $x = 0$, as shown in Figure 5.6. The characteristic picture that emerges is exactly analogous to the piston problem of Section 5.4.1 where now we have to use (5.20) and (5.21) in place of (5.10).

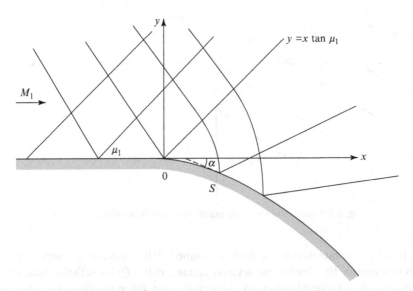

Fig. 5.6 Prandtl–Meyer flow round a smooth corner.

In the incoming flow, $\mu = \mu_1 = \sin^{-1}(1/M_1)$ and $\theta = 0$. Thus, the characteristics are straight lines making angles $\pm\mu_1$ with the x-axis and, by the arguments used in Section 5.4.1, the influence of the corner will not be felt in the uniform region upstream of the characteristic $y = x \tan \mu_1$ through the origin. Again, we can see that the negative characteristics all emanate from the undisturbed region into the simple wave region between the curved wall and $y = x \tan \mu_1$ and, hence,

$$\theta + f(\mu) = f(\mu_1) \tag{5.74}$$

there, where, from (5.20),

$$f(\mu) = \mu + \frac{1}{\lambda} \tan^{-1}(\lambda \cot \mu), \tag{5.75}$$

and, as usual, $\lambda^2 = (\gamma - 1)/(\gamma + 1)$. At a point S on the wall where the slope of the wall is $-\alpha$, we know that $\theta = -\alpha$ and so μ is determined from (5.74). Moreover, from (5.75), we can see that $f(\mu)$ is a decreasing function of μ with a maximum of $\pi/2\lambda$ at $\mu = 0$ and a minimum of $\pi/2$ at $\mu = \pi/2$, as shown in Figure 5.7. Hence, the maximum angle through which the flow can be turned will be $\pi/2\lambda - \pi/2$ and this can only occur when μ_1 is $\pi/2$ and incoming flow is sonic. It is also clear that μ decreases as the flow turns round the corner and, since M increases as μ decreases, this means that as long as our assumption that the incoming flow is supersonic is valid, the flow will stay supersonic throughout. As in Section 5.4.1, we may use the fact that the positive characteristic through S is a straight line along which θ and μ are constant to write down enough equations to determine the flow everywhere (Exercise 5.18).

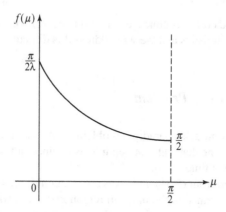

Fig. 5.7 The function $f(\mu)$.

There is one case where we can make easy analytical progress and that is for the supersonic flow past a sharp corner which turns the flow through an angle α as illustrated in Figure 5.8. Now, by analogy with the argument at the end of Section 5.4.1, the flow consists of an expansion fan (in which the positive characteristics are straight lines through the origin) separating a uniform region where $\theta = 0$

and $\mu = \mu_1$ from a second uniform region where $\theta = -\alpha$ and $\mu = \mu_2$, where $f(\mu_2) = \alpha + f(\mu_1)$. The details of the flow in the expansion fan are left to Exercise 5.19.

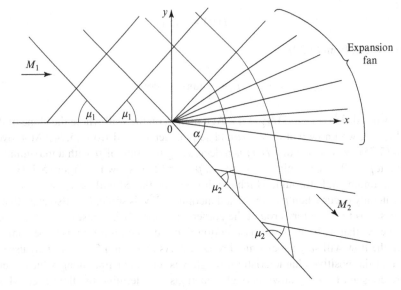

Fig. 5.8 Supersonic flow past a sharp corner.

We note that if the corner is concave, we will once again encounter the problem of intersecting characteristics and we will address this difficulty in Chapter 6.

5.4.3 The Dam Break Problem

For the shallow water model, a paradigm problem to consider is that of a dam breaking suddenly. We assume that water of depth h_0 is retained in $x < 0$ by a dam which is suddenly removed at time $t = 0$.

Using the fact that, from (5.31), $u \pm 2s$ is constant on $dx/dt = u \pm s$, we see that there will be no disturbance in the uniform region $x < -s_0 t$, where $s_0^2 = gh_0$, and that $u + 2s = 2s_0$ everywhere in the simple wave region, as shown in Figure 5.9. Since there is no length scale in this problem, there will be an expansion fan centred on the origin and, along each characteristic through the origin,

$$\frac{x}{t} = u - s.$$

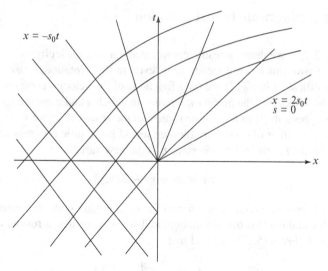

Fig. 5.9 Characteristics for the dam break problem.

Hence,

$$u = \frac{2}{3}\left(\frac{x}{t} + s_0\right) \quad \text{and} \quad s = \frac{1}{3}\left(2s_0 - \frac{x}{t}\right) \tag{5.76}$$

within the fan and, since the depth of the water vanishes when $s = 0$, this solution will only hold for

$$-s_0 t < x < 2s_0 t.$$

The depth of the water at time t is shown in Figure 5.10.

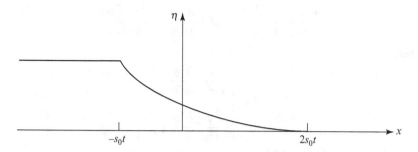

Fig. 5.10 The water depth in the dam break problem.

This solution is analogous to the instantaneous removal of the piston in Section 5.4.1. An obvious extension would be to remove a dam between two reservoirs containing water of different heights, but a quick sketch of the characteristics shows that inevitably some positive characteristics will intersect and so this problem will also be deferred to Chapter 6.

5.5 *The Hodograph Transformation

In Section 5.2.2, we have already commented on the difficulty of solving the equations of two-dimensional steady flow, even in the irrotational case. However, since the coefficients in (5.17) are only functions of the velocity components u and v, the idea of working in the *hodograph plane*, in which u and v are the independent variables, suggests itself as a possible route to a linear model. If we were simply to regard ϕ as a function of these variables we would make little progress and instead we first make the *Legendre transformation* to the new variable

$$\psi(u, v) = xu + yv - \phi(x, y), \tag{5.77}$$

where x, y are now regarded as functions of u, v. There are good geometric reasons (see [43]) for making this transformation, and we will see that it results in a linear equation for ψ. From (5.77), we find that

$$\frac{\partial \psi}{\partial u} = x + \left(u \frac{\partial x}{\partial u} + v \frac{\partial y}{\partial u} - \frac{\partial \phi}{\partial u} \right) = x$$

and similarly, $\partial \psi / \partial v = y$. Then,

$$1 = \frac{\partial}{\partial u} \left(\frac{\partial \phi}{\partial x} \right) = \frac{\partial^2 \phi}{\partial x^2} \frac{\partial^2 \psi}{\partial u^2} + \frac{\partial^2 \phi}{\partial x \partial y} \frac{\partial^2 \psi}{\partial u \partial v}$$

and

$$0 = \frac{\partial}{\partial v} \left(\frac{\partial \phi}{\partial x} \right) = \frac{\partial^2 \phi}{\partial x^2} \frac{\partial^2 \psi}{\partial u \partial v} + \frac{\partial^2 \phi}{\partial x \partial y} \frac{\partial^2 \psi}{\partial v^2}.$$

Hence,

$$\frac{\partial^2 \phi}{\partial x^2} = \frac{1}{D} \frac{\partial^2 \psi}{\partial v^2} \quad \text{and} \quad \frac{\partial^2 \phi}{\partial x \partial y} = -\frac{1}{D} \frac{\partial^2 \psi}{\partial u \partial v},$$

where

$$D = \begin{vmatrix} \dfrac{\partial^2 \psi}{\partial u^2} & \dfrac{\partial^2 \psi}{\partial u \partial v} \\[2ex] \dfrac{\partial^2 \psi}{\partial u \partial v} & \dfrac{\partial^2 \psi}{\partial v^2} \end{vmatrix}.$$

Similarly, differentiation of $\partial \phi / \partial y$ leads to

$$\frac{\partial^2 \phi}{\partial y^2} = \frac{1}{D} \frac{\partial^2 \psi}{\partial u^2}$$

and it can be seen that

$$
\begin{vmatrix} \dfrac{\partial^2 \phi}{\partial x^2} & \dfrac{\partial^2 \phi}{\partial x \partial y} \\[3mm] \dfrac{\partial^2 \phi}{\partial x \partial y} & \dfrac{\partial^2 \phi}{\partial y^2} \end{vmatrix} = \frac{1}{D}.
$$

This transformation is valid as long as D is bounded and non-zero.

Eventually, (5.17) becomes

$$
(c^2 - u^2)\frac{\partial^2 \psi}{\partial v^2} + 2uv\frac{\partial^2 \psi}{\partial u \partial v} + (c^2 - v^2)\frac{\partial^2 \psi}{\partial u^2} = 0, \tag{5.78}
$$

where $c^2 = c_0^2 - [(\gamma - 1)/2](u^2 + v^2)$. This equation is linear and it can be transformed into an even simpler form by writing $u = q\cos\theta$, $v = q\sin\theta$ to get

$$
\frac{\partial^2 \psi}{\partial \theta^2} + \frac{q^2 c^2}{c^2 - q^2}\frac{\partial^2 \psi}{\partial q^2} + q\frac{\partial \psi}{\partial q} = 0, \tag{5.79}
$$

where

$$
c^2 = c_0^2 - \frac{\gamma - 1}{2}q^2;
$$

this is known as *Chaplygin's equation*, and it is clearly susceptible to separation of variables. Chaplygin's equation can be shown to be equivalent to the equations obtained by using the Riemann invariants as independent variables, as described at the end of Section 5.4.1.

Unfortunately, two factors limit the usefulness of (5.79). The first is the condition that D should be bounded and non-zero. In uniform flow, for example, $\psi = 0$ and the flow region maps into a single point on the hodograph plane because $D = 0$. Similarly for simple wave flow, one of the Riemann invariants is constant and the flow region maps into a single characteristic curve in the hodograph plane.

Second, and more crippling, is the fact that everyday boundary conditions in the physical plane can become unmanageable in the hodograph plane. Suppose, for example, there is a fixed boundary on $y = f(x)$; then, the boundary condition is

$$
\frac{\partial \psi}{\partial v} = f(\frac{\partial \psi}{\partial u}) \quad \text{on} \quad v = uf'(\frac{\partial \psi}{\partial u})
$$

and, unless the boundary is straight, the nonlinearity reappears in the problem via the boundary conditions.

Exercises

R 5.1 Suppose that $\partial u/\partial t + u\partial u/\partial x = 0$. Show that if $u(x,0) = u_0(x)$, then $u = u_0(x - ut)$.

When $u_0 = 1/(1 + x^2)$, sketch the evolution of u as t increases by translating the graph of u_0 in the x-direction by a distance that depends on u, and assuming the graph remains smooth. Show that the characteristics are $x = t/(1 + s^2) + s$, where s is a parameter, and that, on the envelope of these characteristics, $2ts = (1 + s^2)^2$. Deduce that $\partial u/\partial x$ first becomes infinite at $x = \sqrt{3}, t = 8\sqrt{3}/9$.

5.2 From (5.14) and (5.16), show that for any function λ,

$$\left[(c^2 - u^2)\frac{\partial}{\partial x} + (\lambda - uv)\frac{\partial}{\partial y}\right]u + \left[(-uv - \lambda)\frac{\partial}{\partial x} + (c^2 - v^2)\frac{\partial}{\partial y}\right]v = 0.$$

Show that the two differential operators in this equation are proportional to each other when $\lambda = \pm c\sqrt{u^2 + v^2 - c^2}$ and deduce (5.18) and (5.19).

Now write $u = cr\cos\theta$, $v = cr\sin\theta$ and show that the slopes of the characteristics are

$$\frac{dy}{dx} = \frac{-r^2 \sin\theta\cos\theta \pm \sqrt{r^2 - 1}}{1 - r^2\cos^2\theta},$$

and then set $r = \operatorname{cosec}\mu$ to derive the equation of the characteristics in the form (5.21).

[If you feel strong you can now go on to check that the Riemann invariants are given by (5.20).]

5.3 Show that if the base of a shallow stream is $z = -b(x)$, then the two-dimensional shallow water equations are

$$\frac{\partial\eta}{\partial t} + \frac{\partial}{\partial x}(u(\eta + b)) = 0,$$

$$\frac{\partial u}{\partial t} + u\frac{\partial u}{\partial x} = -g\frac{\partial\eta}{\partial x}.$$

Deduce that if $b = mx$, then

$$u \pm 2s - mgt = \text{constant on } \frac{dx}{dt} = u \pm s,$$

where $s^2 = g(\eta + mx)$.

Suppose the initial conditions are $u = F(x)$ and $s = s_0 - \frac{1}{2}F(x)$ at $t = 0$. Show that the flow is a simple wave flow with $u + 2s - mgt = 2s_0$ and that

$$u - mgt = F(x - \frac{3}{2}ut + s_0 t + mgt^2).$$

Show that if u and η are small, we retrieve a generalisation of tidal theory (5.33) in the form

$$\frac{\partial^2 u}{\partial t^2} = g \frac{\partial^2}{\partial x^2}(bu).$$

When $b = mx$, this is an equation that changes from elliptic to hyperbolic across $x = 0$, and this gives some insight into how waves break on a sloping beach.

5.4 The scaling used to obtain (5.22)–(5.24) shows that, for two-dimensional waves on shallow water, the equations reduce to

$$\frac{\partial u}{\partial x} + \frac{\partial w}{\partial z} = 0,$$

$$\frac{\partial u}{\partial t} + u\frac{\partial u}{\partial x} + w\frac{\partial u}{\partial z} = -\frac{1}{\rho}\frac{\partial p}{\partial x},$$

$$\frac{\partial p}{\partial z} = -g\rho.$$

Show that the *rotational* flow $u = u_0(z)$, $w = 0$, $\eta = \eta_0 = $ constant satisfies these equations. Show also that, in tidal waves for which $\bar{u} = u - u_0$, w and $\bar{\eta} = \eta - \eta_0$ are small,

$$\frac{\partial \bar{u}}{\partial x} + \frac{\partial w}{\partial z} = 0,$$

$$\frac{\partial \bar{u}}{\partial t} + u_0\frac{\partial \bar{u}}{\partial x} + u_0'w = -g\frac{\partial \bar{\eta}}{\partial x},$$

with $w = 0$ on $z = 0$ and

$$\frac{\partial \bar{\eta}}{\partial t} + u_0\frac{\partial \bar{\eta}}{\partial x} = w \text{ on } z = \eta_0.$$

Now suppose that $\bar{\eta} = \text{Rl}\,(ae^{ik(x-ct)})$ where a, k, c are constant. Show that $w = \text{Rl}\,(f(z)e^{ik(x-ct)})$, where

$$(u_0 - c)f' - u_0'f = igak,$$

with $f(0) = 0, f(\eta_0) = iak(u_0(\eta_0) - c)$. Hence, show that c satisfies

$$g \int_0^{\eta_0} \frac{dz}{(c - u_0(z))^2} = 1.$$

Note that this calculation shows that if there are values of z for which $c = u_0(z)$, then nonlinear terms need to be retained locally near these values of z, which are called *critical layers*.

5.5 Show that for two-dimensional flow, the shallow water equations (5.27) and (5.30) generalise to

$$\frac{\partial u}{\partial t} + u\frac{\partial u}{\partial x} + v\frac{\partial u}{\partial y} = -g\frac{\partial \eta}{\partial x},$$

$$\frac{\partial v}{\partial t} + u\frac{\partial v}{\partial x} + v\frac{\partial v}{\partial y} = -g\frac{\partial \eta}{\partial y},$$

$$\frac{\partial \eta}{\partial t} + \frac{\partial}{\partial x}(u\eta) + \frac{\partial}{\partial y}(v\eta) = 0.$$

Show further that, in steady flow, $\frac{1}{2}(u^2 + v^2) + g\eta$ is conserved along a streamline and that, with $\mathbf{u} = (u, v, 0)$,

$$\mathbf{u} \wedge (\nabla \wedge \mathbf{u}) = \nabla(\frac{1}{2}(u^2 + v^2) + g\eta).$$

Compare this result with Crocco's Theorem (Exercise 2.5), and show that if $\frac{1}{2}(u^2 + v^2) + g\eta$ varies from streamline to streamline, then $\nabla \wedge \mathbf{u} \neq \mathbf{0}$.

*5.6 Three-dimensional waves on shallow water are modelled by

$$\frac{\partial^2 \phi}{\partial z^2} + \delta^2 \left(\frac{\partial^2 \phi}{\partial x^2} + \frac{\partial^2 \phi}{\partial y^2} \right) = 0,$$

with

$$\frac{\partial \phi}{\partial t} + \eta + \frac{\varepsilon}{2\delta^2} \left(\left(\frac{\partial \phi}{\partial z} \right)^2 + \delta^2 \left(\left(\frac{\partial \phi}{\partial x} \right)^2 + \left(\frac{\partial \phi}{\partial y} \right)^2 \right) \right) = 0$$

and

$$\frac{\partial \phi}{\partial z} = \delta^2 \frac{\partial \eta}{\partial t} + \varepsilon\delta^2 \left(\frac{\partial \phi}{\partial x}\frac{\partial \eta}{\partial x} + \frac{\partial \phi}{\partial y}\frac{\partial \eta}{\partial y} \right)$$

on $z = \varepsilon\eta(x, t)$, and

$$\frac{\partial \phi}{\partial z} = 0 \text{ on } z = -1.$$

(i) Suppose that δ and ε are both small and that $\delta^2 \ll O(\varepsilon)$. Suppose also that the waves only vary gradually in the y-direction, with a length scale $\varepsilon^{-\frac{1}{2}}$. Show that over times of $O(\varepsilon^{-1})$, the surface elevation satisfies the dispersionless Kadomtsev–Petviashvili equation

$$\frac{\partial}{\partial \xi} \left(\frac{\partial \eta}{\partial \tau} + \frac{3}{2}\eta\frac{\partial \eta}{\partial \xi} \right) + \frac{1}{2}\frac{\partial^2 \eta}{\partial Y^2} = 0,$$

where $Y = \sqrt{\varepsilon}y$, $\tau = \varepsilon t$, $\xi = x - t$.

(ii) When $\delta^2 = \kappa\varepsilon$ and κ is $O(1)$, deduce that (5.42) and (5.43) generalise to

$$\frac{\partial\phi_0}{\partial t} + \eta_0 + \frac{1}{2}\varepsilon\left(\left(\frac{\partial\phi_0}{\partial x}\right)^2 + \left(\frac{\partial\phi_0}{\partial y}\right)^2\right) + \kappa\varepsilon\left(\frac{\partial A}{\partial t} + \eta_1\right) = O(\varepsilon^2)$$

and

$$\frac{\partial\eta_0}{\partial t} + \frac{\partial^2\phi_0}{\partial x^2} + \frac{\partial^2\phi_0}{\partial y^2} + \varepsilon\left[\eta_0\left(\frac{\partial^2\phi_0}{\partial x^2} + \frac{\partial^2\phi_0}{\partial y^2}\right) + \frac{\partial\phi_0}{\partial x}\cdot\frac{\partial\eta_0}{\partial x} + \frac{\partial\phi_0}{\partial y}\cdot\frac{\partial\eta_0}{\partial y}\right.$$

$$\left. + \kappa\left(\frac{\partial^2}{\partial x^2} + \frac{\partial^2}{\partial y^2}\right)\left(A + \frac{1}{3}\left(\frac{\partial^2\phi_0}{\partial x^2} + \frac{\partial^2\phi}{\partial y^2}\right)\right) + \kappa\frac{\partial\eta_1}{\partial t}\right] = O(\varepsilon^2)$$

, respectively.

Now assume the wave motion is in the x-direction to lowest order, so that ϕ_0 and η_0 are functions of $x - t = \xi$, $\tau = \varepsilon t$ and $Y = \sqrt{\varepsilon}y$. Show that, as in the two-dimensional case,

$$\frac{\partial\phi_0}{\partial\xi} = \eta_0,$$

and hence that the terms of $O(\varepsilon)$ give that ϕ_0 and η_0 also satisfy

$$\frac{\partial\eta_0}{\partial\tau} + \frac{3}{2}\eta_0\frac{\partial\eta_0}{\partial\xi} + \frac{1}{6}\frac{\partial^3\eta_0}{\partial\xi^3} + \frac{1}{2}\frac{\partial^2\phi_0}{\partial Y^2} = 0.$$

This is called the *Kadomtsev–Petviashvili model* (Drazin and Johnson [17]).

*5.7 Show that if $\phi(x, t)$ satisfies (5.34) with $\phi(x, 0) = \phi_0(x)$, $-\infty < x < \infty$, then its Fourier transform $\bar{\phi} = \int_{-\infty}^{\infty}\phi(x, t)e^{ikx}dx$ is

$$\bar{\phi}(k, t) = \bar{\phi}_0(k)e^{(ik-\varepsilon k^2)t}.$$

Deduce that, as $\varepsilon \to 0$ for $x, t = O(1)$,

$$\phi = \frac{1}{2\pi}\int_{-\infty}^{\infty}\bar{\phi}_0(k)e^{ik(t-x)-\varepsilon k^2 t}dk \sim \phi_0(x - t).$$

Show that when $t = \tau/\varepsilon$, where $\tau = O(1)$ and $\xi = x - t + O(1)$,

$$\phi = \frac{e^{-\xi^2/4\tau}}{2\pi}\int_{-\infty}^{\infty}\bar{\phi}_0(k)e^{-\tau(k+\frac{i\xi}{2\tau})^2}dk.$$

When $\bar{\phi}_0(k)$ is well behaved at $k = -i\xi/2\tau$, show that

$$\phi \sim \frac{\text{constant}}{\sqrt{\tau}} e^{-\xi^2/4\tau}$$

as $\tau \to \infty$. To what initial condition for (5.35) does this solution correspond?

*5.8 Set $\varepsilon = 1$ in (5.37)–(5.39) and show that, as $\delta \to 0$

$$\frac{\partial \phi_0}{\partial t} + \eta_0 + \frac{1}{2}\left(\frac{\partial \phi_0}{\partial x}\right)^2 = 0$$

and

$$\frac{\partial^2 \phi_0}{\partial x^2} + \frac{\partial \eta_0}{\partial t} + \frac{\partial \phi_0}{\partial x} \cdot \frac{\partial \eta_0}{\partial x} + \eta_0 \frac{\partial^2 \phi_0}{\partial x^2} = 0,$$

using the notation of (5.40). Show that these equations are equivalent to the shallow water equations (5.27) and (5.30).

Note that in the derivation of (5.27) and (5.30), no assumption was made about irrotationality and yet the above method relies on (5.36), which assumes irrotationality. From (5.28), we see that $\boldsymbol{\omega} = \nabla \wedge \mathbf{u} = (0, -z\partial^2 u/\partial x^2, 0)$ and, using the scalings of (5.36), $|\boldsymbol{\omega}| = O(\delta^2 \sqrt{g/h})$. Hence, in the shallow water approximation, the vorticity may be taken to be zero.

*5.9 Immiscible fluids of density ρ_w, ρ_0 (perhaps water and oil) flow along a horizontal channel $0 < y < D$ with the oil above the water. The interface is at $y = h(x, t)$. Making the shallow water assumptions in each fluid, show that the horizontal velocities u_w and u_0 satisfy

$$\frac{\partial h}{\partial t} + \frac{\partial}{\partial x}(hu_w) = 0,$$

$$-\frac{\partial h}{\partial t} + \frac{\partial}{\partial x}((D - h)u_0) = 0,$$

$$\frac{\partial u_w}{\partial t} + u_w \frac{\partial u_w}{\partial x} + g\frac{\partial h}{\partial x} + \frac{1}{\rho_w}\frac{\partial p}{\partial x} = 0$$

and

$$\frac{\partial u_0}{\partial t} + u_0 \frac{\partial u_0}{\partial x} + g\frac{\partial h}{\partial x} + \frac{1}{\rho_0}\frac{\partial p}{\partial x} = 0,$$

where $p(x, t)$ is the pressure at the interface. Show that for small disturbances about the uniform state $h = H$, $u_w = u_0 = U$, $p = p_0$, the perturbation $\tilde{h} = h - H$ satisfies

$$\left(\frac{\rho_0}{D-H} + \frac{\rho_w}{H}\right)\left(\frac{\partial^2 \bar{h}}{\partial t^2} + 2U\frac{\partial^2 \bar{h}}{\partial x \partial t} + U^2\frac{\partial^2 \bar{h}}{\partial x^2}\right) + g(\rho_0 - \rho_w)\frac{\partial^2 \bar{h}}{\partial x^2} = 0.$$

Deduce that waves can propagate on the interface as long as $\rho_0 < \rho_w$.

*5.10 With $A = re^{i\theta}$ in (5.53) show that

$$r = r_0 = \text{constant},$$

$$\theta = -\frac{5}{3}r_0^2 T + \text{constant}.$$

Deduce that the effect of the nonlinearity over timescales of $t = O(\varepsilon^{-2})$ is to change the period of the solution (whose amplitude is given by $r = r_0$) from 2π (as predicted by (5.51)) to $2\pi/(1 - \frac{5}{3}\varepsilon^2 r_0^2)$.
Confirm that the solution is, in fact, periodic over all time scales when $r_0 = O(1)$ by sketching the phase plane of (5.47), for which the phase curves are

$$y^2 + x^2 = \frac{2}{3}\varepsilon x^3 + \text{constant},$$

where $y = dx/dt$.

*5.11 Show that in axes $\xi = x - c_0 t$ moving with the speed of sound in a stationary gas, the equation for the one-dimensional flow of a heat conducting gas are

$$\frac{\partial \rho}{\partial t} + (u - c_0)\frac{\partial \rho}{\partial \xi} + \rho\frac{\partial u}{\partial \xi} = 0,$$

$$\rho\frac{\partial u}{\partial t} + \rho(u - c_0)\frac{\partial u}{\partial \xi} + \frac{\partial p}{\partial \xi} = 0$$

and

$$\frac{\partial p}{\partial t} + (u - c_0)\frac{\partial p}{\partial \xi} + \gamma p\frac{\partial u}{\partial \xi} = \frac{kR}{c_v}\frac{\partial^2 T}{\partial \xi^2},$$

in the usual notation. To study the long-time evolution of a small-amplitude wave, write $t = \varepsilon^{-1}\tau$, $u \sim \varepsilon u_1 + \varepsilon^2 u_2 + \cdots$, $p \sim p_0 + \varepsilon p_1 + \varepsilon^2 p_2 + \cdots$, $\rho \sim \rho_0 + \varepsilon \rho_1 + \varepsilon^2 \rho_2 + \cdots$, $T \sim T_0 + \varepsilon T_1 + \cdots$, where $p_0 = \rho_0 R T_0$. Then, assuming $kR/c_v = \varepsilon \bar{k}$, derive the system

$$-c_0\frac{\partial \rho_1}{\partial \xi} + \rho_0\frac{\partial u_1}{\partial \xi} + \varepsilon\left[\frac{\partial \rho_1}{\partial \tau} + u_1\frac{\partial \rho_1}{\partial \xi} + \rho_1\frac{\partial u_1}{\partial \xi} - c_0\frac{\partial \rho_2}{\partial \xi} + \rho_0\frac{\partial u_2}{\partial \xi}\right] = O(\varepsilon^2),$$

$$-c_0\rho_0\frac{\partial u_1}{\partial \xi} + \frac{\partial p_1}{\partial \xi}$$
$$+ \varepsilon\left[\rho_0\frac{\partial u_1}{\partial \tau} + \rho_0 u_1\frac{\partial u_1}{\partial \xi} - c_0\rho_1\frac{\partial u_1}{\partial \xi} - c_0\rho_0\frac{\partial u_2}{\partial \xi} + \frac{\partial p_2}{\partial \xi}\right] = O(\varepsilon^2),$$

$$-c_0\frac{\partial p_1}{\partial \xi} + \gamma p_0\frac{\partial u_1}{\partial \xi}$$
$$+ \varepsilon\left[\frac{\partial p_1}{\partial \tau} + u_1\frac{\partial p_1}{\partial \xi} + \gamma p_1\frac{\partial u_1}{\partial \xi} - c_0\frac{\partial p_2}{\partial \xi} + \gamma p_0\frac{\partial u_2}{\partial \xi} - \bar{k}\frac{\partial^2 T_1}{\partial \xi^2}\right] = O(\varepsilon^2).$$

Deduce that $\rho_1 = (\rho_0/c_0)u_1$, $p_1 = \rho_0 c_0 u_1$, $p_1 = R(\rho_0 T_1 + T_0\rho_1)$, and hence, combine these equations and eliminate p_1, ρ_1 to obtain

$$2\rho_0 c_0\frac{\partial u_1}{\partial \tau} + (\gamma + 1)\rho_0 c_0 u_1\frac{\partial u_1}{\partial \xi} = \bar{k}\frac{\partial^2 T_1}{\partial \xi^2} = \frac{\bar{k}T_0}{c_0}(\gamma - 1)\frac{\partial^2 u_1}{\partial \xi^2}.$$

This equation can easily be reduced to *Burgers equation* (see later, viz (6.11)). Unfortunately, the assumption that thermal conduction can be retained while viscosity is ignored is not true for a gas like air, but the effect of viscosity is merely to change the coefficients in Burgers equation.

*5.12 (i) Take $\delta = 1$ and show that (5.37) and (5.38) can then be written as

$$\frac{\partial \phi}{\partial t} + \eta + \varepsilon\left[\eta\frac{\partial^2 \phi}{\partial z\partial t} + \frac{1}{2}\left(\frac{\partial \phi}{\partial x}\right)^2 + \frac{1}{2}\left(\frac{\partial \phi}{\partial z}\right)^2\right] = O(\varepsilon^2)$$

and

$$\frac{\partial \phi}{\partial z} - \frac{\partial \eta}{\partial t} + \varepsilon\left[\eta\frac{\partial^2 \phi}{\partial z^2} - \frac{\partial \phi}{\partial x}\frac{\partial \eta}{\partial x}\right] = O(\varepsilon^2)$$

on $z = 0$. Deduce that, to lowest order, Q in (5.55) is given by

$$-2\left(\frac{\partial \phi}{\partial x}\frac{\partial^2 \phi}{\partial x\partial t} + \frac{\partial \phi}{\partial z}\frac{\partial^2 \phi}{\partial z\partial t}\right)$$

in accordance with (5.58).

(ii) Show that, when the terms on the left-hand side of (5.55) are expanded in terms of the variables given before (5.57), the term of $O(\varepsilon^2)$ will be

$$\varepsilon^2\left(\frac{\partial^2 \phi_2}{\partial t^2} + \frac{\partial \phi_2}{\partial z} + 2\frac{\partial^2 \phi_1}{\partial t\partial T_1} + \frac{\partial^2 \phi_0}{\partial T_1^2} + 2\frac{\partial^2 \phi_0}{\partial t\partial T_2}\right),$$

all evaluated on $z = 0$. Use the formulae for ϕ_0, ϕ_1 and ϕ_2 between (5.56) and (5.59) to show that the terms in this expression that are proportional to $e^{i(kx-\omega t)}$ are

$$\left(-i\frac{\partial A}{\partial X_2} - i\frac{\partial B}{\partial \xi} + i\frac{\partial B}{\partial \xi} + V^2\frac{\partial^2 A}{\partial \xi^2} - 2i\omega\frac{\partial A}{\partial T_2} \right) e^{i(kx-\omega t)}.$$

Unfortunately, there are 19 nonlinear terms of $O(\varepsilon^2)$ which are proportional to $e^{i(kx-\omega t)}$, but, following the pattern of (i), it is straightforward (but tedious) to see that they are all cubic in ϕ_0 and involve two t-derivatives and three x- or z-derivatives. Hence, they are all proportional to $\omega^2 k^3 A^2 A^* e^{i(kx-\omega t)}$. Use this information to deduce (5.59).

*5.13 Show that, to the lowest order in ε, writing $\phi = -x\sin t + \tilde{\phi}$ in equations (5.65) to (5.69) leads to

$$\nabla^2 \tilde{\phi} = 0,$$

with

$$\frac{\partial \tilde{\phi}}{\partial x} = 0 \quad \text{on } x = 0, \pi,$$

$$\frac{\partial \tilde{\phi}}{\partial z} = 0 \quad \text{on } z = -h$$

and

$$\frac{\partial^2 \tilde{\phi}}{\partial t^2} + \frac{g}{a\omega^2}\frac{\partial \tilde{\phi}}{\partial z} = -x\sin t \text{ on } z = 0.$$

If $g/a\omega^2 = (1 + \delta)\coth h$, show that writing

$$\tilde{\phi} = \left(\sum A_n \cos nx \cosh n(z + h) \right) \sin t$$

leads to a solution in which

$$A_m \left(\cosh nh - n(1 + \delta)\coth h \sinh nh \right) = \frac{2}{\pi n^2}\left((-1)^n - 1 \right).$$

Hence, show that as $\delta \to 0$

$$\tilde{\phi} \sim \frac{4\cos x \cosh(z + h)\sin t}{\pi\delta\cosh h} + O(1).$$

*5.14 In the notation of Section 5.3.2, show that writing

$$A \sim e^{i\Omega\tau}(B_0 + \varepsilon B_1 + \cdots)$$

in (5.60), where $B_0 = \sqrt{(-\Omega/c)}$, leads to the equation

$$i\frac{\partial B_1}{\partial \tau} - V^3\frac{\partial^2 B_1}{\partial \xi^2} - \Omega(B_1 + B_1^*) = 0.$$

Now write

$$B_1 = B_{10}e^{i(k\xi-\sigma\tau)} + B_{11}e^{-i(k\xi-\sigma\tau)},$$

where B_{10}, B_{11} are constants and show that

$$\sigma^2 = V^3k^2(V^3k^2 - 2\Omega).$$

Deduce that the wave train given by (5.56) with $A = \sqrt{(-\Omega/c)}e^{-i\Omega\tau}$ is unstable for all $\Omega > 0$.

R 5.15 The equations

$$\left(\frac{\partial}{\partial t} + (u \pm c)\frac{\partial}{\partial x}\right)\left(u \pm \frac{2c}{\gamma - 1}\right) = 0$$

for the gas velocity u and sound speed c are used to model gas flow in a tube under the action of a piston at $x = X(t)$. The gas is in $x < X(t)$ and $X(0) = 0$ and $\dot{X}(t) \geq 0$, $\ddot{X}(t) \geq 0$. When the gas is initially at rest with $c = c_0$, show that

$$u = \dot{X}(\tau),$$

where

$$u + \frac{2c}{\gamma - 1} = \frac{2c_0}{\gamma - 1} \quad \text{and} \quad \frac{x - X(\tau)}{t - \tau} = u - c$$

in the region $-c_0t < x < X(t)$. Deduce that

(i) when $\dot{X}$ is a constant greater than $2c_0/(\gamma - 1)$, the gas expands into the region

$$-c_0 < \frac{x}{t} < \frac{2c_0}{\gamma - 1};$$

(ii) when $X = gt^2/2$, then

$$\gamma u = \left(c_0 + \frac{\gamma + 1}{2}gt\right) - \left[\left(c_0 + \frac{\gamma + 1}{2}gt\right)^2 - 2\gamma g(c_0t + x)\right]^{1/2}$$

in the region $-c_0 < \frac{x}{t} < \frac{gt}{2}$, for $t < \frac{2c_0}{(\gamma - 1)g}$.

R 5.16 Suppose that in the piston problem in Exercise 5.15 above, the piston path is $x = -\frac{1}{2}gt^2$, where $g > 0$. Show that the negative characteristics are

$$\frac{x + \frac{1}{2}g\tau^2}{t - \tau} = -c_0 - \frac{\gamma + 1}{2}g\tau,$$

and deduce that, for small τ, these characteristics form an envelope at

$$x = -c_0 t, \quad t = 2c_0/(\gamma + 1)g.$$

Verify that $\partial u/\partial x$ is infinite at this point.

R 5.17 A piston is withdrawn impulsively from a tube containing gas in $x < 0$. The model of Section 5.2 is

$$\left(\frac{\partial}{\partial t} + (u \pm c)\frac{\partial}{\partial x}\right)\left(u \pm \frac{2c}{\gamma - 1}\right) = 0$$

with $u = 0$, $c = c_0 = $ constant for $t = 0$, $x < 0$, and $u = V$ on $x = Vt$ for $t > 0$, with $V > 0$.

Taking L to be an *arbitrary* length scale, non-dimensionalise this model by writing

$$x = Lx', \quad t = (L/c_0)t', \quad u = Vu', \quad c = c_0 c'$$

to give

$$\left(\frac{\partial}{\partial t'} + (Mu' \pm c')\frac{\partial}{\partial x'}\right)\left(Mu' \pm \frac{2c'}{\gamma - 1}\right) = 0,$$

where $M = V/c_0$, and

$$u' = 0, \quad c' = 1 \quad \text{for } t' = 0, \quad x' = 0,$$

$$u' = 1 \quad \text{on } x' = t' \quad \text{for } t' > 0.$$

Now use the fact that u' is evidently only a function of x', t' and M to deduce that u/V is only a function of x/Vt and M.

Deduce that the solution is

$$u = \begin{cases} 0, & \text{if } \dfrac{x}{Vt} < -\dfrac{1}{M} \\[2mm] \dfrac{2V}{\gamma + 1}\left(\dfrac{x}{Vt} + \dfrac{1}{M}\right) & \text{if } -\dfrac{1}{M} < \dfrac{x}{Vt} < \dfrac{\gamma + 1}{2} - \dfrac{1}{M} \\[2mm] V, & \text{if } \dfrac{\gamma + 1}{2} - \dfrac{1}{M} < \dfrac{x}{Vt} < 1. \end{cases} \tag{*}$$

Confirm this result by using (5.72) and (5.73) and noting that only small values of τ are relevant in generating the expansion fan. Use this idea to show that

$$x = t\left[\frac{\gamma + 1}{2}\dot{X}(\tau) - c_0\right],$$

and hence retrieve the solution (*).

R 5.18 With reference to Figure 5.6, show that on any negative characteristic,

$$\theta + f(\mu) = f(\mu_1),$$

where f is defined by (5.75). Show also that, at the point S where the flow deflection is $-\alpha$, μ is given by

$$f(\mu) = f(\mu_1) + \alpha.$$

Show further that, on the positive characteristic through S,

$$\theta - f(\mu) = -2\alpha - f(\mu_1),$$

and infer that θ and μ are both constant on this characteristic.
Writing the boundary as $y = -F(x)$, show that the solution at the point (x, y) in the simple wave region is

$$\theta(x, y) = -\alpha, \quad f(\mu(x, y)) = f(\mu_1) + \alpha,$$

where $\alpha = \tan^{-1} F'(\xi)$ and $(y + F(\xi))/(x - \xi) = \theta + \mu$.
Now use Figure 5.7 to show that $\mu \to 0$ when α increases to $\pi/2\lambda - f(\mu_1)$, and hence that the greatest angle through which a flow can be turned is $\pi/2(1/\lambda - 1)$.

5.19 Specialise the answer to Exercise 5.18 to the case when $F(x) = x \tan \alpha$ to show that, in the simple wave region, θ and μ are given by

$$\tan(\theta + \mu) = y/x, \quad \theta + f(\mu) = f(\mu_1).$$

How could you show that θ and μ are only functions of y/x without solving the equations?
Show that the simple wave region is

$$\tan \mu_1 > y/x > \tan(-\alpha + f^{-1}(\alpha + f(\mu_1))).$$

5.20 Fluid of depth s_0^2/g is contained in a tank $-1 < x < 0$, and, at time $t = 0$, the right-hand wall is moved in a positive direction with speed $U(< 2s_0)$, while the left-hand wall is held fixed at $x = -1$. Assuming that the shallow water equations are valid in the subsequent flow, draw a characteristic diagram and show that

$$u = \frac{2}{3}\left(\frac{x}{t} + s_0\right)$$

in a region bounded by $x + s_0 t = 0$, $x + (s_0 - 3U/2)t = 0$ and $x = 2s_0 t - 3(s_0 t)^{1/3}$.

*5.21 Gas at rest with speed of sound c_0 is contained between diaphragms at $x = \pm a$ in a long tube. At $t = 0$ the diaphragms are broken and the gas

flows along the tube in both directions into a vacuum. Sketch the characteristics of the subsequent flow in the (x, t) plane. Show that in $x > 0$, there is
an expansion fan (simple wave flow) in which the solution is

$$u = \frac{2}{\gamma + 1} \left(\frac{x - a}{t} + c_0 \right), \quad c = \frac{\gamma - 1}{\gamma + 1} \left(\frac{2c_0}{\gamma - 1} - \frac{x - a}{t} \right)$$

and that this region is bounded by the curves

$$x = a - c_0 t, \quad x = a + \frac{2c_0}{\gamma - 1} t, \quad x = a + \frac{2c_0}{\gamma - 1} t - 2v c_0^{1 - 1/v} a^{1/v} t^{1 - 1/v},$$

where $v = (\gamma + 1)/(2(\gamma - 1))$.

In order to solve the problem when the expansion fans interact, we need to
change to new variables. By writing $2r = u + 2c/(\gamma - 1)$ and
$2s = -u + 2c/(\gamma - 1)$, show that the equations of the characteristics (5.10)
reduce to

$$\frac{\partial x}{\partial s} = \left(\frac{\gamma + 1}{2} r - \frac{3 - \gamma}{2} s \right) \frac{\partial t}{\partial s}$$

and

$$\frac{\partial x}{\partial t} = \left(\frac{3 - \gamma}{2} r - \frac{(\gamma + 1)}{2} s \right) \frac{\partial t}{\partial r}.$$

Hence, show that the equation for t as a function of r, s is

$$\frac{\partial^2 t}{\partial r \partial s} + \frac{v}{(r + s)} \left(\frac{\partial t}{\partial r} + \frac{\partial t}{\partial s} \right) = 0, \qquad (**)$$

which is called an *Euler–Poisson–Darboux equation* (see [11]). The equation (**) is the one-dimensional unsteady gas dynamics analogue of Chaplygin's equation (5.79) for two-dimensional steady flow.
Show that the boundary conditions on

$$x = \pm \left(a + \frac{2c_0 t}{\gamma - 1} - 2v c_0 t \left(\frac{a}{c_0 t} \right)^{1/v} \right)$$

transform to

$$t = \frac{a}{c_0} \left[\frac{2c_0}{(\gamma - 1)(s + c_0/(\gamma - 1))} \right]^v \quad \text{on } r = c_0/(\gamma - 1)$$

and

$$t = \frac{a}{c_0} \left[\frac{2c_0}{(\gamma - 1)(r + c_0/(\gamma - 1))} \right]^v \quad \text{on } s = c_0/(\gamma - 1).$$

It is possible to solve for $t(r, s)$ explicitly in terms of hypergeometric functions by using the Riemann function technique (see Garabedian [20]). However, it is amusing to note that if we set $r + s = \xi, r - s = \eta$, equation (**) becomes

$$\frac{\partial^2 t}{\partial \xi^2} + \frac{2\nu}{\xi} \frac{\partial t}{\partial \xi} = \frac{\partial^2 t}{\partial \eta^2},$$

which is just the wave equation in $2\nu + 1$ dimensions if ξ is identified with the radial direction and η is identified with time. Therefore, we can use the general solutions of Exercise 4.27 whenever $2\nu + 1$ is an integer. Taking $2\nu + 1 = n$ implies that $\gamma = n/(n-2)$, so, for air with $\gamma = 7/5$, we need to solve the wave equation in 7 dimensions!

5.22 Show that in unsteady one-dimensional gas flow, the continuity equation

$$\frac{\partial \rho}{\partial t} + \frac{\partial}{\partial x}(\rho u) = 0$$

implies the existence of a function ξ such that

$$\frac{\partial \xi}{\partial x} = \rho, \quad \frac{\partial \xi}{\partial t} = -\rho u.$$

Deduce that, with ξ, t as independent (Lagrangian) variables,

$$\frac{\partial x}{\partial \xi} = \frac{1}{\rho}, \quad \frac{\partial^2 x}{\partial t^2} = -\frac{\partial p}{\partial \xi} \quad \text{and} \quad \frac{\partial}{\partial t}\left(\frac{p}{\rho^\gamma}\right) = 0.$$

5.23 Suppose that a shallow layer of water flows down an inclined plane that makes a small angle α with the horizontal. Show that, if the x-axis is along the line of greatest slope and the y-axis is perpendicular to the plane, then two-dimensional shallow water flow is governed by the equations

$$\frac{\partial u}{\partial x} + \frac{\partial v}{\partial y} = 0,$$

$$\frac{\partial u}{\partial t} + u\frac{\partial u}{\partial x} + v\frac{\partial u}{\partial y} = -\frac{1}{\rho}\frac{\partial p}{\partial x} + g\alpha,$$

$$0 = -\frac{1}{\rho}\frac{\partial p}{\partial y} - g,$$

if α is suitably small. If the surface of the water is given by $y = \eta(x, t)$, show that $p = -\rho g(y - \eta)$ and that on $y = \eta$, $v = \partial\eta/\partial t + u\partial\eta/\partial x$. Deduce that

$$\frac{\partial \eta}{\partial t} + \frac{\partial}{\partial x}(u\eta) = 0$$

and

$$\frac{\partial u}{\partial t} + u\frac{\partial u}{\partial x} + g\frac{\partial \eta}{\partial x} = g\alpha,$$

and show that, in steady flow with $u\eta = m$,

$$\frac{1}{2}\frac{m^2}{\eta^2} + g\eta - g\alpha x = \text{constant}.$$

Show also that there is a possible solution with η constant in which the flow has constant acceleration $g\alpha$ down the plane.

Chapter 6
Shock Waves

6.1 Discontinuous Solutions

The time has come to face up to the task of making a mathematical model that can deal with flows containing *shock waves* or *shocks*, across which the various dependent physical variables themselves have discontinuities. Such discontinuities are often called *jump discontinuities* in contrast to situations in which only the derivatives of the physical variables have discontinuities.

We have already been motivated to study such shock waves in our study of both resonance in Section 4.2 and nozzle flow in Section 4.7.2. In neither case were we able to find a physically acceptable smooth solution and we were thus led to postulate the possibility of jump discontinuities. Even more compelling, however, was the analysis of Section 5.4 where we saw clear evidence that nonlinear wave propagation frequently leads to a breakdown in the continuity of the flow variables. Not only did we find that smooth solutions could fail to exist if discontinuities were imposed in either the boundary or the initial data as, for example, in the impulsively started piston problem of Section 5.4.1 but also, more interestingly, we have seen that discontinuities could arise spontaneously in certain flows in which the data is arbitrarily smooth.

Our theory will apply almost exclusively to systems of partial differential equations that can be written in the form

$$\frac{\partial \mathbf{P}}{\partial t} + \sum_i \frac{\partial \mathbf{Q}_i}{\partial x_i} = \mathbf{R},$$

where $\mathbf{P}$, $\mathbf{Q}_i$ and $\mathbf{R}$ are functions only of the dependent variable $\mathbf{u}$ and the independent variables. Such systems are called systems of *conservation laws*, and we have already seen many examples. Indeed, in our basic gas dynamic model, (2.6) of Chapter 2 is already such a law for mass conservation, and (2.7) and (2.9) are consequences of conservation laws for momentum and energy.

© Springer Science+Business Media New York 2015
H. Ockendon, J.R. Ockendon, *Waves and Compressible Flow*,
Texts in Applied Mathematics 47, DOI 10.1007/978-1-4939-3381-5_6

6.1.1 Introduction to Weak Solutions

The problem of finding a discontinuous solution to a model which is posed as a system of partial differential equations may be clarified by consideration of a paradigm involving just a single variable.

We suppose that in the one-dimensional flow of a "continuum" of density $\rho(x, t)$, the mass flux, ρu, is a *prescribed* function $f(\rho)$. An example of such a situation would be the simplest "continuum" model for traffic flow (Whitham [54]) and the resulting solution is often called a *kinematic wave*, as already encountered in Section 5.1.

For a continuous flow, the equation of conservation of mass (2.6) will be

$$\frac{\partial \rho}{\partial t} + \frac{\partial}{\partial x} f(\rho) = 0, \tag{6.1}$$

and we have seen already in Section 5.1 that such equations can readily admit multivalued solutions. We suppose that the single-valued solution to this problem has a discontinuity at $x = X_s(t)$, where ρ jumps from ρ_1 to ρ_2 as shown in Figure 6.1. (Here, and henceforth, shocks are denoted by double lines).

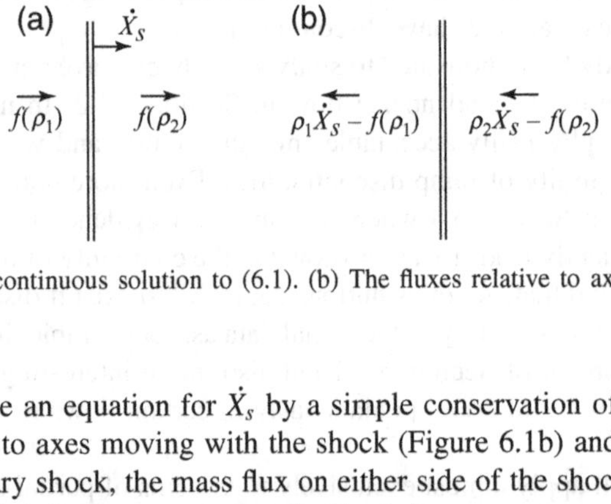

Fig. 6.1 (a) A discontinuous solution to (6.1). (b) The fluxes relative to axes moving with the shock.

We can derive an equation for $\dot{X}_s$ by a simple conservation of mass argument. We first change to axes moving with the shock (Figure 6.1b) and then, relative to the now-stationary shock, the mass flux on either side of the shock is $f(\rho_i) - \rho_i \dot{X}_s$. Thus, the mass flux discrepancy across the shock is

$$[f(\rho) - \rho \dot{X}_s],$$

where, as usual, the square brackets denote the size of the jump of the enclosed quantity. Thus, if we prohibit any sources or sinks of mass,

$$[f(\rho)] = [\rho] \dot{X}_s, \tag{6.2}$$

and this physically derived law will hold across any such shock. We might hope that this *jump condition* (6.2) might be sufficient to determine the position of a shock uniquely, but we will see that this is not necessarily the case. However, before we study the uniqueness question more closely, we first think more generally about how we might have derived the condition (6.2).

The key idea is to generalise the *derivation* of the equation of conservation of mass, so that no assumption about the differentiability of the dependent variables is needed. For our one-dimensional continuum, the mass in the interval $a(t) < x < b(t)$ is

$$m(t) = \int_{a(t)}^{b(t)} \rho \, dx,$$

where $a(t)$ and $b(t)$ are any functions of t. The mass flux into (a, b) at time t will be

$$q(t) = [f(\rho) - \rho \dot{a}]_{x=a(t)} - [f(\rho) - \rho \dot{b}]_{x=b(t)}, \tag{6.3}$$

and, in the time interval $(t, t + \delta t)$, the mass balance for the region $a < x < b$ can be written as

$$\delta m = m(t + \delta t) - m(t) = q(t)\delta t + O(\delta t^2).$$

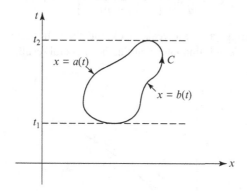

Fig. 6.2 The contour C in the (x, t) plane.

Now, if the curves $x = a(t)$ and $x = b(t)$ form a closed curve C for $t_1 < t < t_2$ as shown in Figure 6.2, we can integrate (6.3) to obtain[1]

$$\int_{t_1}^{t_2} q(t) \, dt = \oint_C f(\rho) \, dt - \rho \, dx. \tag{6.4}$$

But $q = dm/dt$ and hence the integral on the left-hand side of (6.4) vanishes. Since this formula is true for any closed curve C in the (x, t) plane, we can use the divergence theorem to show that this relation is equivalent to (6.1) for flows in which ρ is a differentiable function. On the other hand, if ρ has a discontinuity on the shock $x = X_s(t)$, then, by taking C to be the "pillbox" contour $A_1 A_2 B_2 B_1 A_1$ shown in

---

[1] If the interior of C is not a convex region, we may have to divide it up into convex regions in order to do the integration illustrated in Figure 6.3, but the final result (6.4) will still hold.

Figure 6.3, we can see that as $A_1, A_2 \to A$ and $B_1, B_2 \to B$, so that $A_1 B_1$ and $A_2 B_2$ lie along $x = X_s(t)$ on opposite sides of the shock, (6.4) gives

$$\int_{A_1}^{B_1} \rho \, dx - f(\rho) \, dt = \int_{A_2}^{B_2} \rho \, dx - f(\rho) \, dt.$$

This may be written as

$$\int_A^B [\rho] \, dx - [f(\rho)] \, dt = 0,$$

and since AB is *any* segment of the curve $x = X_s(t)$,

$$[\rho] \, dx = [f(\rho)] \, dt$$

on the shock and (6.2) follows immediately.

Note that as $[\rho] \to 0$, equation (6.2) gives

$$\dot{X}_s = \lim_{\rho_2 \to \rho_1} \left\{ \frac{f(\rho_2) - f(\rho_1)}{\rho_2 - \rho_1} \right\} = f'(\rho_1),$$

and so, in the limiting case of a weak shock, the shock lies along the characteristic of equation (6.1). We will find later that this is a general result.

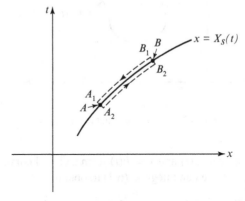

Fig. 6.3 The contour $A_1 A_2 B_1 B_2 A_1$ for the discontinuity at $x = X_s(t)$.

This approach, which uses the integral formulation (6.4) of the problem, means that we can cater for discontinuous as well as continuous solutions and these more general solutions are known as *weak solutions*. For our purposes, a weak solution (this adjective must not be confused with its use in the previous paragraph!) is a solution that is piecewise smooth and which satisfies (6.2) at any points of discontinuity. There is an extensive theory for weak solutions of partial differential equations which are conservation laws, in which the above arguments are made rigorous by the use of *test functions*.[2] However, because the models we consider here are motivated by physical considerations, we can appeal directly to arguments such as those used above.

We now summarise the ideas above before applying them to more general situations. If a conservation law can be written in the integral form

$$\oint_C P\,dx - Q\,dt = 0, \tag{6.5}$$

where P and Q are functions of x and t and C is any smooth closed curve in the domain in which the solution is sought, then a differentiable solution will satisfy the partial differential equation

$$\frac{\partial P}{\partial t} + \frac{\partial Q}{\partial x} = 0, \tag{6.6}$$

and, across shock on $x = X_s(t)$,

$$\frac{dx}{dt} = \frac{[Q]}{[P]}. \tag{6.7}$$

We still have to be very careful when using physical arguments that we have the physically acceptable form for the integral formulation (6.5). For example, (6.1) could equally well be written as

$$\frac{\partial}{\partial t}\left(\frac{1}{2}\rho^2\right) + \frac{\partial}{\partial x}F(\rho) = 0,$$

where $F'(\rho) = \rho f'(\rho)$, and this would correspond to an integral formulation

$$\int \frac{1}{2}\rho^2\,dx - F(\rho)\,dt = 0,$$

and a jump condition

$$\left[\frac{1}{2}\rho^2\right]\dot{X}_s(t) = [F(\rho)],$$

which is quite different from (6.2). Thus, there may be a number of integral formulations corresponding to the same basic partial differential equation, each of which will give rise to a different jump condition.

Even when we have decided on the correct jump conditions, it may still be possible to find a number of possible discontinuous solutions. For instance, let us consider (6.1) and the jump condition (6.2) with $f(\rho) = \frac{1}{2}\rho^2$ and initial conditions

---

[2] The basic idea of a weak solution to (6.1) is to replace (6.4) by $\iint_S (f\partial\phi/\partial x + \rho\partial\phi/\partial t)\,dx dt = 0$, where S is an arbitrary fixed region in the (x, t) plane and ϕ is *any* suitably smooth test function (see [43]).

$$\rho(x,0) = \begin{cases} 0, & x < 0 \\ 1, & x > 0. \end{cases}$$

It is easy to see that three possible weak solutions are

$$\text{(i)}\ \rho(x,t) = \begin{cases} 0, & x < \tfrac{1}{2}t \\ 1, & x > \tfrac{1}{2}t, \end{cases} \tag{6.8}$$

$$\text{(ii)}\ \rho(x,t) = \begin{cases} 0, & x < \tfrac{1}{4}t \\ \tfrac{1}{2}, & \tfrac{1}{4}t < x < \tfrac{3}{4}t \\ 1, & x > \tfrac{3}{4}t \end{cases} \tag{6.9}$$

and

$$\text{(iii)}\ \rho(x,t) = \begin{cases} 0, & x < 0 \\ \tfrac{x}{t}, & 0 < x < t \\ 1, & x > t, \end{cases} \tag{6.10}$$

and we can construct many more. In crude terms, the generalisation involved in formulating the problem as (6.5) is dangerous; it allows for discontinuous solutions but, at the moment, it allows far too many of them. Thus, we will need to appeal to some extra information in order to decide which of the many possible weak solutions is relevant to the physical situation.

This problem of nonuniqueness may be resolved in three distinct ways, as discussed in [43]. The first method is to use specific physical arguments, the second is to use general thermodynamic principles and the third is to use the principle of *causality*, which basically asserts that the future cannot influence the past. We will return to the latter two ideas later in the chapter and just employ the first method here.

We suppose that our continuum model (6.1) is generalised to allow for small diffusional effects. Thus, whenever ρ varies spatially, we assume that there will be a small diffusion flux, proportional to $-\partial\rho/\partial x$, which transports material from higher to lower densities. This introduces a second-order derivative into (6.1) so that, in the case when $f = \tfrac{1}{2}\rho^2$, we are once more led to consider Burgers Equation

$$\frac{\partial \rho}{\partial t} + \rho \frac{\partial \rho}{\partial x} = \varepsilon \frac{\partial^2 \rho}{\partial x^2}, \tag{6.11}$$

where ε is a small *positive* constant.

If we suppose that a wave $\rho(x,t)$ advances into ambient material $\rho = \rho_1$ as $x \to +\infty$, it is reasonable to seek a *travelling wave* in which $\rho = \rho(x - Vt)$ and $V > 0$ is constant. It is easy to see that no interesting smooth travelling wave can exist when $\varepsilon = 0$, but, for $\varepsilon > 0$, we obtain

$$\varepsilon \rho' = \frac{1}{2}\rho^2 - V\rho + \text{constant},$$

where the prime denotes differentiation with respect to $x - Vt$. Thus, if $\rho \to \rho_2$ as $x - Vt \to -\infty$,

$$\varepsilon\rho' = \frac{1}{2}(\rho - \rho_1)(\rho - \rho_2)$$

and $V = \frac{1}{2}(\rho_1 + \rho_2)$ which, with $V = \dot{X}_s$, is just jump condition (6.2)! Moreover, by sketching the solutions in the $(\rho, x - Vt)$ plane we can see that a solution which allows a transition from ρ_1 to ρ_2 when $\varepsilon > 0$ is only possible if $\rho' < 0$ and $\rho_1 \leq \rho_2$. The width of this transition region is of $O(\varepsilon)$ and the fact that, even as $\varepsilon \to 0$, the density *increases* as the wave passes, leads us to the "selection principle" that we need if we are to ensure that there is just one physically acceptable solution satisfying the jump condition (6.2). When this principle is applied to the example above, we find that all the discontinuous solutions such as (6.8) and (6.9) violate this condition, and the only acceptable solution is the continuous "expansion" given by (6.10). Note the importance of the sign of the diffusion coefficient ε and the choice of $f(\rho) = \frac{1}{2}\rho^2$ in this argument; had ε been negative or if f had been a cubic in ρ, say, we would have been led to a different selection principle.

We now have the beginnings of a systematic theory for (6.1) with $f = \frac{1}{2}\rho^2$. This theory eventually leads to the result that, on the whole line $-\infty < x < \infty$, all weak solutions tend, as $t \to \infty$, to a combination of *N-waves*, in which spatially linear segments of the form $(x + x_0)/(t + t_0)$, where x_0 and t_0 are constants, are separated by jump discontinuities satisfying (6.2) (see Lax [34]).

It can be shown (see Exercise 6.6) that the introduction of weak viscosity into the equations of gas dynamics leads systematically to (6.11) as the model for the pressure or density in a shock wave. Hence, the requirement that the shock is *compressive*, so that the gas pressure increases as the shock passes, is indeed the appropriate selection principle for a physically acceptable gas dynamic shock.

The idea of weak solutions can be generalised to flows in more than one space dimension. Suppose, for example, that in two dimensions the conservation law for a differentiable flow is

$$\frac{\partial \rho}{\partial t} + \frac{\partial}{\partial x}f(\rho) + \frac{\partial}{\partial y}g(\rho) = 0; \tag{6.12}$$

then, (6.4) is replaced by

$$\int_S (\rho, f, g) . \mathbf{dS} = 0, \tag{6.13}$$

where S is any smooth closed surface in (t, x, y) space. Writing S as $F(t, x, y) = 0$, so that

$$\mathbf{dS} = \left(\frac{\partial F}{\partial t}, \frac{\partial F}{\partial x}, \frac{\partial F}{\partial y}\right)\left(\left(\frac{\partial F}{\partial t}\right)^2 + \left(\frac{\partial F}{\partial x}\right)^2 + \left(\frac{\partial F}{\partial y}\right)^2\right)^{-1/2} dS,$$

and noting that the normal velocity v_n of a point on S is

$$v_n = \frac{-\partial F/\partial t}{\left((\partial F/\partial x)^2 + (\partial F/\partial y)^2\right)^{1/2}},$$

then, an argument analogous to that used to derive (6.2) leads to the jump condition

$$[\rho]v_n = [(f, g).\mathbf{n}],\tag{6.14}$$

where $\mathbf{n} = (\partial F/\partial x, \partial F/\partial y)/((\partial F/\partial x)^2 + (\partial F/\partial y)^2)^{1/2}$ is the normal to the projection of S in the x, y plane at time t.

We can remark here that the free surfaces in our gravity wave models in Chapter 3 can be regarded as jumps from $\rho = 0$ on one side of the boundary (the air) to $\rho =$ constant in the water. Hence, corresponding to the continuity equation (2.6), the jump condition derived from (6.14) leads directly to the free boundary condition (3.10).

6.1.2 Rankine–Hugoniot Shock Conditions

We now consider one-dimensional unsteady gas flow with shocks. Motivated by the previous section, we start by writing down integral formulations for conservation of mass, momentum and energy.

We can use (6.4) directly to write down conservation of mass in the form

$$\oint_C \rho \, dx - \rho u \, dt = 0,\tag{6.15}$$

and thus, from (6.7), across a shock

$$\dot{X}_s = \frac{[\rho u]}{[\rho]}.\tag{6.16}$$

In a similar way, the conservation of momentum is written as

$$\oint_C \rho u \, dx - (\rho u^2 + p) \, dt = 0,\tag{6.17}$$

to give the shock condition

$$\dot{X}_s = \frac{[\rho u^2 + p]}{[\rho u]}.\tag{6.18}$$

Finally, the energy equation is

$$\oint_C \left(\frac{1}{2}\rho u^2 + \rho e\right) dx - \left(\frac{1}{2}\rho u^3 + \rho e u + p u\right) dt = 0,\tag{6.19}$$

and, on putting $e = c_v T = p/(\gamma - 1)\rho$, this leads to the third shock condition

$$\dot{X}_s = \frac{[\frac{1}{2}\rho u^3 + \gamma p u/(\gamma - 1)]}{[\frac{1}{2}\rho u^2 + p/(\gamma - 1)]}.\tag{6.20}$$

Note that in the same way that we were able to go from (6.5) to (6.6), we can easily derive the equations for one-dimensional flow without shocks from the integral formulations (6.15), (6.17) and (6.19) and, after some manipulation, we arrive at equations (5.3)–(5.5) as expected. However, we emphasise that the shock relation (6.20) must be derived from the *conservation* form of the energy equation, and we note that, in spite of (5.5), p/ρ^γ will *not* be conserved across a shock.

The three shock relations or jump conditions given by (6.16), (6.18) and (6.20) are called the *Rankine–Hugoniot relations* and are usually written in the form

$$[\rho(u - \dot{X}_s)] = 0, \tag{6.21}$$

$$[p + \rho(u - \dot{X}_s)^2] = 0 \tag{6.22}$$

and

$$\left[\frac{\gamma p}{(\gamma - 1)\rho} + \frac{1}{2}(u - \dot{X}_s)^2\right] = 0. \tag{6.23}$$

Equation (6.21) comes directly from (6.16), but some algebraic manipulation is needed to obtain (6.22) and (6.23) (see Exercise 6.1).

It can be seen from (6.21)–(6.23) that it is the velocity *relative* to the shock that appears naturally in the shock relations. This is not surprising since, as we saw in Section 6.1.1, physically motivated jump conditions are most conveniently written down by considering the flow *relative to the shock*.

Further properties of shocks can be understood by rewriting jump conditions in terms of M_1, the upstream Mach number relative to the shock. We use suffix 1 for upstream variables ahead of the shock, suffix 2 for downstream variables behind the shock, and put $M_i = (\dot{X}_s - u_i)/c_i$. Then, as shown in Exercise 6.2, the downstream variables may be expressed in terms of the upstream ones as

$$\frac{p_2}{p_1} = \frac{2\gamma M_1^2}{\gamma + 1} - \frac{\gamma - 1}{\gamma + 1}, \tag{6.24}$$

$$\frac{\rho_2}{\rho_1} = \frac{\dot{X}_s - u_1}{\dot{X}_s - u_2} = \frac{(\gamma + 1)M_1^2}{2 + (\gamma - 1)M_1^2} \tag{6.25}$$

and

$$M_2^2 = \frac{(\gamma - 1)M_1^2 + 2}{2\gamma M_1^2 - (\gamma - 1)}. \tag{6.26}$$

Now, following the clue from the end of Section 6.1.1, we demand that the shock should be *compressive* so that

$$p_2 \geq p_1. \tag{6.27}$$

Observations reveal that this inequality is satisfied in practice by all purely gas dynamic shock waves, and it implies from (6.24)–(6.26) that

$$M_1 \geq 1, \quad M_2 \leq 1 \quad \text{and} \quad \rho_2 \geq \rho_1. \tag{6.28}$$

Thus, relative to the shock, the flow upstream (ahead) must be supersonic, while the flow downstream (behind) is subsonic. A less immediate implication of (6.27) is that

$$\frac{p_2}{\rho_2^\gamma} \geq \frac{p_1}{\rho_1^\gamma}, \tag{6.29}$$

which shows that the entropy of the gas always *increases* as it passes through the shock (Exercise 6.3). Indeed, were we to be guided by thermodynamics, we could appeal to the Second Law to assert (6.29) and then deduce (6.27). The equivalence of this approach to the one we have adopted is unsurprising since the dissipation inherent in asserting that $dQ/dt \geq 0$ in (2.20) is also inherent in asserting that the coefficient of viscosity, which is analogous to the diffusion coefficient ε in (6.11), is positive. Note also that increase in entropy across a shock depends on the shock speed $\dot{X}_s$ and this means that we can never have homentropic flow downstream of a genuinely unsteady shock for which $\ddot{X}_s \neq 0$.

We now illustrate these results with an example. We reconsider the piston problem discussed in Section 5.4.1, but now we suppose that the piston is pushed *into* the gas in $x > 0$ with constant speed V. Let us look for a solution in which a compressive shock travels into the quiescent gas with constant speed $\dot{X}_s$ so that, behind this shock, the gas moves with the piston at constant speed V. We assume the pressure and density ahead of the shock are p_0 and ρ_0 and behind the shock are p_1 and ρ_1. Then, writing down the three Rankine–Hugoniot equations allows us to determine the three unknown quantities $\dot{X}_s$, p_1 and ρ_1. Eliminating p_1 and ρ_1, we obtain the equation

$$\dot{X}_s^{\ 2} - \frac{(\gamma + 1)}{2} V \dot{X}_s - c_0^2 = 0 \tag{6.30}$$

for $\dot{X}_s$, where $c_0^2 = \gamma p_0 / \rho_0$. Since $\dot{X}_s$ must be positive,

$$\dot{X}_s = \frac{\gamma + 1}{4} V + \left[\frac{(\gamma + 1)^2 V^2}{16} + c_0^2 \right]^{1/2},$$

and the characteristics are as shown in Figure 6.4. The slope of the shock is *between* the slopes of the upstream and downstream positive characteristics and this configuration, which is typical of all evolutionary shocks, can be shown to be a manifestation of the *principle of causality*, which is described in detail in [43]. In the limiting case, as $V \to 0$ and the shock becomes very weak, the shock can again be seen to lie along the positive characteristic $x = c_0 t$.

6.1.3 Shocks in Two-dimensional Steady Flow

It is quite easy to generalise our results for a one-dimensional steady shock with $\dot{X}_s = 0$ to a straight two-dimensional shock simply by imposing a velocity parallel to the shock on the whole system. However, we proceed more systematically by

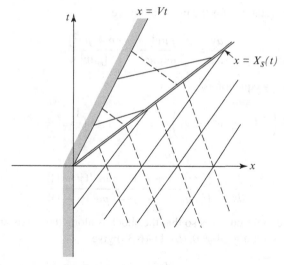

Fig. 6.4 The compressive piston problem: —— positive characteristics, - - - negative characteristics.

deriving the shock relations *ab initio* by using the method of weak solutions for two-dimensional steady flow.

The integral equations of motion in this case are even easier to write down than those of the previous section. Conservation of mass of a fluid in steady flow implies that

$$\oint_C \rho \mathbf{u.n} \, ds = 0$$

around any contour C, and so

$$\oint \rho u \, dy - \rho v \, dx = 0.$$

Hence, from (6.7), the jump condition is

$$\frac{dy}{dx} = \frac{[\rho v]}{[\rho u]}. \tag{6.31}$$

There are two components of the momentum equation and, in the x-and y-directions, we obtain

$$\oint_C (p + \rho u^2) \, dy - \rho u v \, dx = 0$$

and

$$\oint_C \rho \, uv \, dy - (p + \rho v^2) \, dx = 0.$$

Thus, the shock relations for the momentum are

$$\frac{dy}{dx} = \frac{[\rho uv]}{[p + \rho u^2]} = \frac{[p + \rho v^2]}{[\rho uv]}. \tag{6.32}$$

Finally, the energy equation is

$$\oint_C \left(pu + \rho u \left(\frac{1}{2}(u^2 + v^2) + e \right) \right) dy - \left(pv + \rho v (\frac{1}{2}(u^2 + v^2) + e) \right) dx = 0,$$

so that

$$\frac{dy}{dx} = \frac{[\frac{1}{2}\rho v(u^2 + v^2) + \gamma pv/(\gamma - 1)]}{[\frac{1}{2}\rho u(u^2 + v^2) + \gamma pu/(\gamma - 1)]}. \tag{6.33}$$

Let us now choose our axes so that the shock is along the y-axis and so dy/dx is infinite. Then, assuming $[p] \neq 0$, (6.31)–(6.33) give

$$[\rho u] = 0, \tag{6.34}$$

$$[p + \rho u^2] = 0, \tag{6.35}$$

$$[v] = 0 \tag{6.36}$$

and

$$\left[\frac{1}{2}u^2 + \gamma p/(\gamma - 1)\rho \right] = 0 \tag{6.37}$$

which are, as expected, the Rankine–Hugoniot equations normal to a steady shock as derived in Section 6.1.2, with the extra condition that the velocity parallel to the shock is conserved.

Since the condition (6.27) that the shock be compressive implies that the velocity of the gas normal to the shock is decreased as it passes through a shock, we can see immediately from Figure 6.5 that the effect of the shock is to turn the flow *towards* the shock. Note also that it is only the component of the velocity normal to the shock that must be supersonic ahead of the shock and subsonic behind it. Thus, although the overall flow must be supersonic ahead of the shock, so that $M_1 > 1$, we only know that $M_2 \sin(\beta - \theta) < 1$. Hence, the flow downstream could be either subsonic or supersonic. Indeed, transonic aeroplanes have swept back wings partly to lessen the possibility of shocks forming at the leading edges of the wings.

We remark that (6.31)–(6.33) are all direct consequences of the unsteady shock relation (6.14). We may also expect the relations (6.34)–(6.37) to hold across any smoothly evolving shock, straight or curved, as long as the jumps are taken with respect to axes moving with the shock and aligned so that the x-axis is in the direction of the normal to the shock.

It is often convenient to work with the actual velocities before and after a stationary oblique shock rather than with the components along and perpendicular to

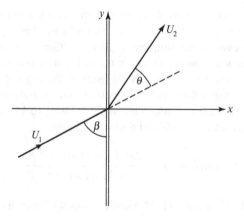

Fig. 6.5 Flow deflection due to an oblique shock.

the shock. We therefore rewrite (6.34)–(6.37) in terms of U_1 and U_2, as indicated in Figure 6.5, to get

$$\rho_1 U_1 \sin \beta = \rho_2 U_2 \sin(\beta - \theta), \tag{6.38}$$

$$p_1 + \rho_1 U_1^2 \sin^2 \beta = p_2 + \rho_2 U_2^2 \sin^2(\beta - \theta), \tag{6.39}$$

$$U_1 \cos \beta = U_2 \cos(\beta - \theta) \tag{6.40}$$

and

$$\frac{1}{2} U_1^2 \sin^2 \beta + \frac{\gamma p_1}{(\gamma - 1)\rho_1} = \frac{1}{2} U_2^2 \sin^2(\beta - \theta) + \frac{\gamma p_2}{(\gamma - 1)\rho_2}. \tag{6.41}$$

We note that, using (6.40), (6.41) can be written as

$$\frac{1}{2} U_1^2 + \frac{\gamma p_1}{(\gamma - 1)\rho_1} = \frac{1}{2} U_2^2 + \frac{\gamma p_2}{(\gamma - 1)\rho_2}. \tag{6.42}$$

Now we can once again manipulate these equations in terms of M_1 and β so that, instead of (6.24)–(6.26), we have

$$\frac{p_2}{p_1} = \frac{2\gamma M_1^2 \sin^2 \beta}{\gamma + 1} - \frac{\gamma - 1}{\gamma + 1}, \tag{6.43}$$

$$\frac{\rho_2}{\rho_1} = \frac{\tan \beta}{\tan(\beta - \theta)} = \frac{(\gamma + 1)M_1^2 \sin^2 \beta}{2 + (\gamma - 1)M_1^2 \sin^2 \beta} \tag{6.44}$$

and

$$M_2^2 \sin^2(\beta - \theta) = \frac{(\gamma - 1)M_1^2 \sin^2 \beta + 2}{2\gamma M_1^2 \sin^2 \beta - (\gamma - 1)}. \tag{6.45}$$

Although these shock relations, together with the condition of compression or entropy increase across the shock, appear straightforward enough, alarming complexities can arise even in the simplest application. Consider, for example, supersonic flow into a concave corner as illustrated in Figure 6.6 in which the flow is to be turned through an angle θ. This is analogous to the ingoing piston problem in one-dimensional unsteady flow, and we now try to find a straight shock in the (x, y) plane which will turn the flow through the angle θ. Using (6.44), we can deduce that the shock angle β is related to M_1 and θ via the formula

$$\tan(\beta - \theta) = \frac{2 + (\gamma - 1) M_1^2 \sin^2 \beta}{(\gamma + 1) M_1^2 \sin \beta \cos \beta}. \tag{6.46}$$

Note that as $\theta \to 0$, $\beta \to \sin^{-1}(1/M_1)$ and the shock is again a characteristic.

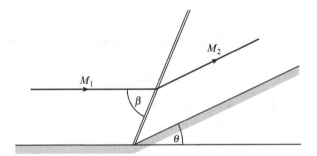

Fig. 6.6 Supersonic flow past a concave corner.

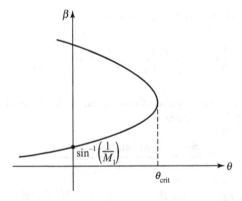

Fig. 6.7 The shock angle β as a function of the deflection θ.

However, when we plot β against θ in Figure 6.7, we find that there are two possible values for β for each $\theta < \theta_{crit}$.[3] Now it can be argued (but the details are beyond the scope of this book) that the upper branch, which represents the stronger shock, is unstable, but, in order to decide what happens if $\theta > \theta_{crit}$, we need some experimental evidence. It turns out that the shock is no longer straight when $\theta > \theta_{crit}$ and that it also "stands off" from the wedge as shown in Figure 6.8. Now we have lost the simple situation of two uniform flows each with constant entropy, and, since behind the curved shock the flow is no longer homentropic, the full equations of gas dynamics will need to be solved there. Note that the shock will be normal to the incoming flow at A and thus the flow immediately behind the shock there is always subsonic.[4]

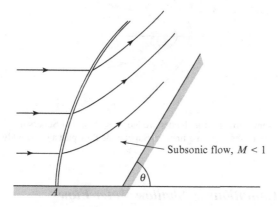

Fig. 6.8 Supersonic flow past a concave corner when $\theta > \theta_{crit}$.

These ideas may be used to understand the supersonic flow past a two-dimensional wing with a sharp leading edge (Figure 6.9(a) and 6.9(b)) and this leads us ultimately to the even more complicated problem of supersonic flow past a *blunt body* as shown in Figure 6.9(c). The blunt body clearly creates a curved shock with stand off, and, if the body is also slender enough, one can expect that, although there will be a subsonic region near the nose, the flow will rapidly accelerate to become supersonic downstream of the *sonic line* on which $M = 1$. This change from subsonic to supersonic flow makes the numerical problem particularly troublesome. However, there are limiting situations that can be studied analytically as we shall see in Sections 6.2.2 and 6.3.1.

---

[3] An alternative way of describing the flow through an oblique shock graphically is to use the *shock polar* as in Exercise 6.8.

[4] In view of the dimensionality arguments of Section 5.4.2 and Exercise 5.19, this phenomenon should not occur for a corner in an infinite wall. However, because the flow behind the shock at A is subsonic, the equations are locally elliptic, and the solution will depend on conditions downstream that will, in practice, involve a length scale which determines the stand-off distance and the shock curvature. We will see another example of such a configuration at the end of Section 6.2.2.

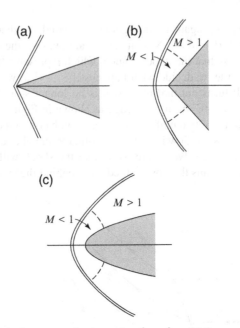

Fig. 6.9 (a) Supersonic flow past a slender wedge $\theta < \theta_{crit}$. (b) Supersonic flow past a wedge with $\theta > \theta_{crit}$; - - - indicates the sonic line. (c) Supersonic flow past a blunt body; - - - indicates the sonic line.

6.1.4 Jump Conditions in Shallow Water Flow

We motivate our discussion of discontinuous solutions of the shallow water equations by making two practical observations. The first is that *bores*, which are steep waves separating regions of different but nearly constant water depth, can be driven upstream in estuarine rivers at periods of high tide. Secondly, the radial jet formed when the jet from a kitchen tap impinges on a flat-bottomed sink usually suffers an abrupt increase in depth at a definite distance from the point of impingement. In both cases, the sudden jump is reminiscent of a shock wave; the water flows from a fast shallow region, which can be observed to be supercritical, into a slower deeper region, which can be observed to be subcritical, and we will find a close analogy between such water flows and supersonic and subsonic gas flows.

Since the shallow water equations (5.27) and (5.30) are only approximations to the conservation form of the Euler equations, it is dangerous to try to formulate the equations in integral form directly. Hence, we study these "hydraulic" discontinuities using physical concepts to motivate the appropriate Rankine–Hugoniot relations for shallow water.

Using the notation of Section 5.2.3, conservation of mass across a bore moving with speed $\dot{X}_s$ will immediately imply

$$[\eta(u - \dot{X}_s)] = 0. \tag{6.47}$$

The other equation of shallow water flow, (5.27), can be thought of as either conservation of momentum or conservation of energy but, in a continuous flow, both these quantities are conserved by this one equation, as was the case for the inviscid incompressible Euler equation (2.7). Now, because we only have two first-order equations (5.27) and (5.30), there can be only two Rankine–Hugoniot conditions and hence, across a bore, it is possible to conserve *either* momentum *or* energy but not both. In most situations, ranging from the kitchen sink example to a bore on a river, it is more realistic to conserve momentum, and energy will be dissipated. For weak bores, this energy is mainly transported away from the discontinuity by a train of waves and this is known as an *undular bore*, whereas for a stronger *turbulent bore* the energy is mostly dissipated in the form of turbulence at a wave crest.[5] Either type of bore may also be referred to as a *hydraulic jump*, although this term is more usually reserved for a stationary discontinuity.

It can also happen that momentum can be destroyed by a sluice gate or a small obstacle on the bed of the river. In this case, a smooth transition can be created for which it may be more appropriate to conserve energy rather than momentum. However, for the rest of this chapter we will study momentum-conserving bores.

First we make use of the fact that the change in momentum of the flow on either side of the bore must be balanced by the pressure forces acting. Thus,

$$[\rho\eta(u - \dot{X}_s)u] + \left[\int_0^\eta (p - p_0)\, dz\right] = 0,$$

and, using (6.47) and remembering that $p - p_0 = \rho g(\eta - z)$, we obtain

$$\left[\rho\eta(u - \dot{X}_s)^2 + \frac{1}{2}\rho g\eta^2\right] = 0. \tag{6.48}$$

Using the now-familiar conservation of mass argument, and rewriting (6.47) and (6.48) in terms of s, where $s^2 = g\eta$, we obtain the shock relations in the form

$$[s^2(u - \dot{X}_s)] = 0 \tag{6.49}$$

and

$$[s^2(u - \dot{X}_s)^2 + \frac{1}{2}s^4] = 0 \tag{6.50}$$

Note that although the partial differential equations for unsteady one-dimensional gas dynamics can be translated into those for shallow water by putting $c = s$ and $\gamma = 2$, the Rankine–Hugoniot shock relations (6.21)–(6.23) are nothing like (6.49) and (6.50) above. Hence, although hydraulic tanks can be used in the laboratory to simulate continuous homentropic gas flow, they cannot be used to study gas flows with shocks. However, we remark that we could have obtained (6.49), (6.50) from the gas dynamic shock conditions (6.21), (6.22) for a gas with $\gamma = 2$ had we assumed that the entropy, and hence p/ρ^2, was conserved across the shock. We

---

[5] Practical observations indicate that the transition from an undular bore to a turbulent bore occurs as the ratio of the increase in depth to the original depth increases through a value of around 0.3.

know that entropy is not conserved across a shock in a real gas, but a good model for genuine shocks in water (not hydraulic jumps) is to use the Rankine–Hugoniot conditions (6.21), (6.22) together with conservation of entropy $[p/\rho^\gamma] = 0$, with $\gamma \simeq 7$, as noted in the footnote on p. 11.

We cannot ignore energy altogether and indeed, energy considerations are vital when it comes to selecting physically acceptable solutions of (6.49) and (6.50). We must make sure that energy is dissipated rather than gained across a momentum-preserving discontinuity, and, when we do this, we derive a condition which is analogous to that derived from the compressive condition (6.27) for a gas dynamic shock. As a discontinuity converts fluid from velocity u_1, depth η_1 to velocity u_2, depth η_2, the rate of increase in kinetic and potential energy is

$$
\Delta \dot{E}_1 = \left[\frac{1}{2} \rho \eta u^2 (\dot{X}_s - u) + \int_0^\eta \rho g z (\dot{X}_s - u)\, dz \right]_1^2
$$
$$
= \frac{\rho}{2g} [(u^2 + s^2) s^2 (\dot{X}_s - u)]_1^2.
$$

However, energy is also created by the work done by the pressure forces, and the rate at which this work is done is

$$
\Delta \dot{E}_2 = \left[\int_0^\eta (p - p_0) u\, dz \right]_2^1
$$
$$
= -\frac{\rho}{2g} [s^4 u]_1^2.
$$

Thus, on using (6.49) and (6.50), the total rate at which energy is gained is

$$
\Delta \dot{E}_1 + \Delta \dot{E}_2 = \frac{\rho}{g} s_1^2 (\dot{X}_s - u_1) \left[\frac{1}{2} (u - \dot{X}_s)^2 + s^2 \right]_1^2, \tag{6.51}
$$

which can be rewritten as

$$
-\frac{\rho (\dot{X}_s - u_1)(s_2^2 - s_1^2)^3}{4 g s_2^2} \tag{6.52}
$$

after further manipulation. Thus, we see immediately that energy cannot be conserved. Moreover, since energy must be lost, we see that, if $\dot{X}_s > u_1$, then $s_2 > s_1$. Hence, for a momentum-conserving turbulent bore on shallow water, the flow in front of the discontinuity will be shallower than the flow behind it. In the same way, in a stationary hydraulic jump, the flow can only jump from fast shallow flow to slower deeper flow. With a little more work, it can be shown (Exercise 6.11) that, relative to the bore, the flow ahead of the discontinuity is supercritical and the flow behind is subcritical; this is exactly analogous to the result for a plane shock in a gas where, relative to the shock, the flow ahead is supersonic and the flow behind is subsonic.

As an example, we consider a piston being pushed with constant velocity V into static water of depth s_1^2/g. If we assume that a bore runs ahead of the piston with constant speed $\dot{X}_s$ and that the depth of the water behind the bore is s_2^2/g, we can use (6.49) and (6.50) to get

$$s_2^2(V - \dot{X}_s) = -s_1^2\dot{X}_s \tag{6.53}$$

and

$$s_2^2(V - \dot{X}_s)^2 + \frac{1}{2}s_2^4 = s_1^2\dot{X}_s^{\ 2} + \frac{1}{2}s_1^4. \tag{6.54}$$

Hence, eliminating s_2, the equation for $\dot{X}_s$ is

$$\dot{X}_s^{\ 3} - 2V\dot{X}_s^{\ 2} + \dot{X}_s(V^2 - s_1^2) + \frac{1}{2}s_1^2V = 0,$$

which is very different from (6.30), the equation for the shock speed in the analogous gas dynamic problem.

6.1.5 *Delta Shocks in Shallow Water

So far our approach has relied on modelling shock waves as generalised weak solutions of nonlinear hyperbolic partial differential equations. This has enabled us to bypass the fact that none of our hyperbolic models is physically realistic when the dependent variables have jump discontinuities. It is therefore gratifying that, when we resort to arguments involving the addition of physical mechanisms such as viscosity, we are led to the same Rankine–Hugoniot conditions.

Shallow water theory motivates the study of an even more extreme class of singular solutions in the limit when the Froude number, $U/\sqrt{gh}$, becomes very large. To illustrate this we consider a piston, with speed V, moving fast into shallow water of depth s_1^2/g and consider equations (6.53) and (6.54) in the limit as $V/s_1 \to \infty$. We find that

$$\frac{\dot{X}_s - V}{V} \sim \frac{s_1}{\sqrt{2}V} \ll 1 \text{ and } \frac{s_2^2}{s_1^2} \sim \frac{\sqrt{2}V}{s_1} \gg 1,$$

so that the speed of the shock approaches the piston speed and the jump in depth tends to infinity as $V/s_1 \to \infty$. In particular, if the piston starts to move impulsively at $t = 0$, a narrow region of deep fluid, in which the depth is almost a delta function, will form ahead of the piston.

Clearly a configuration like this makes our shallow water model even less physically realistic than it is for a hydraulic jump in which the derivatives of s and u contain delta functions. Yet, from the mathematical point of view, we can proceed by considering even "weaker" generalised solutions of the shallow water model in which s itself can involve delta functions. This leads to the theory of *delta shocks* which, at a very basic level, is equivalent to writing the velocity and depth in the shallow water equations as linear combinations of delta functions (corresponding to delta shocks), step functions (corresponding to hydraulic jumps or shocks) and smooth functions [33, 35]. We also remark that, in the high Froude number limit, the two families of characteristics described in Section 5.2.3 become nearly parallel. Thus, in the limit, the model is no longer strictly hyperbolic, which gives another warning that a dramatic change in behaviour is about to occur. Further details can be found in [19].

6.2 Other Flows involving Shock Waves

In this section, we consider a number of more complicated flows in which shocks can arise. First we consider some gas dynamic examples and then one from shallow water theory, all the examples being extensions of cases that have been considered earlier in this chapter.

6.2.1 Shock Tubes

We are now in a position to generalise the instantaneously removed piston problem of Section 5.4.1 to the more physically realistic case of the sudden rupture of a diaphragm separating high and low pressure gas in a *shock tube*. Immediately after the rupture, the low pressure gas will be compressed by the high pressure gas and, at first sight, we are led to conjecture the scenario sketched in Figure 6.10, where a shock propagates into the low pressure gas on the right and an expansion wave propagates into the high pressure gas on the left. Thus, we can use the solution in Section 5.4.1 for the expansion wave and the solution from Section 6.1.2 for the shock wave generated by an ingoing piston. However, when we try to piece together these two solutions, we find that it is not possible to construct a solution which is continuous away from the shock. We need to introduce a new kind of discontinuity at the position where the low pressure gas which has traversed the shock meets the expanded high pressure gas from the left of the diaphragm. In general, these two regions of gas will have different entropies and so it will be impossible to make both the pressure and the density continuous at this point. If we go back to the Rankine–Hugoniot shock conditions (6.16), (6.18), (6.20), we can see that it is possible to have discontinuities in which

$$[u] = 0, \quad [p] = 0, \quad [\rho] \neq 0, \tag{6.55}$$

as long as $u = \dot{X}_s$ on both sides of the discontinuity. This special solution of the Rankine–Hugoniot equation is called a *contact discontinuity* and, in this problem, a contact discontinuity will travel with the gas particles which were originally adjacent to the diaphragm. Hence, the final scenario illustrated in Figure 6.10 must contain a contact discontinuity along the line OC, as shown in detail in Exercise 6.17. Note that in the limiting case where there is a vacuum in the right-hand half of the tube initially, we retrieve the solution mentioned in Section 5.4.1, when the piston is removed instantaneously.

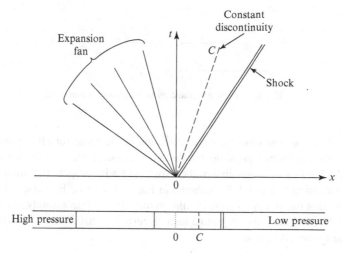

Fig. 6.10 A shock tube.

From a mathematical point of view, it is always helpful to think of jump discontinuities in the limit as they become vanishingly weak. We have already seen in Section 6.1.2 that, as the jumps tend to zero, the shock tends to either the positive or negative characteristic. However, when we take the same limit for a contact discontinuity, its path remains the particle path $dX_s/dt = u$, which is the third characteristic of the system of equations (5.3)–(5.5).

A similar analysis can be performed for a sudden dambreak in which there is originally water at different levels on both sides of the dam. A bore will travel into the shallower water and an expansion wave will travel into the deeper water, but in this case, there will be *no* contact discontinuity since the original system of equations (5.27) and (5.30) has just two characteristics (Exercise 6.14).

6.2.2 Oblique Shock Interactions

We can use the Rankine–Hugoniot conditions for an oblique shock to construct another composite flow, this time in two dimensions. Suppose we consider the flow generated by two corners as shown in Figure 6.11 and ask what happens downstream of the shock interaction.

Since, from (5.21), we know that the characteristics make angles $\pm\sin^{-1}(1/M)$ with the flow direction and that shocks become characteristics when they are weak, it is tempting to postulate, at least for small flow deflections, the existence of two crossing "transmitted" shocks as shown in Figure 6.12. These shocks must

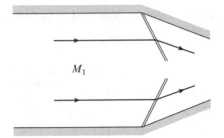

Fig. 6.11 Flow generated by two concave corners.

be arranged so that the ultimate flow deflection is the same for all incoming fluid particles. This is indeed possible for certain parameter regimes of the variables M_1, β_1, β_2, but it is only possible at the expense of admitting a "two-dimensional" contact discontinuity along OP, as shown in Figure 6.12 (see Exercise 6.18 for details). Note that the velocity perpendicular to the contact discontinuity must be zero on both sides, and the jump in p must also be zero, but that from (6.31)–(6.33), the discontinuity can allow both

$$[\rho] \neq 0 \text{ and } [v] \neq 0,$$

where v is the component of velocity parallel to the discontinuity. Thus, this discontinuity is a vortex sheet, and we recall from Section 4.4.2 that a vortex sheet is always subject to the Kelvin–Helmholtz instability. Hence, we expect it to rapidly "smear out" into a turbulent layer that mixes the gas on either side of the sheet.

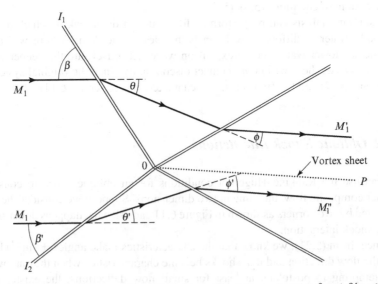

Fig. 6.12 Shock interaction; case 1: incident shocks I_1 and I_2 are at angles β and β' to the free stream.

However, this is not the end of the story because it is sometimes possible for two "incoming" shocks to generate just one outgoing shock and a contact discontinuity, as shown in Figure 6.13. When the lower shock is normal to the oncoming stream, this is the so-called *Mach reflection* phenomenon, which is often observed when shocks interact with viscous boundary layers (Liepmann and Roshko [36]). This leads us to consider more generally the reflection of a plane shock wave that is incident at an angle β_1 on a plane wall, the Mach number upstream of the shock being M_1. One possibility is the straightforward scenario shown in Figure 6.14 in which the incident shock gives rise to a single plane reflected shock. This is certainly

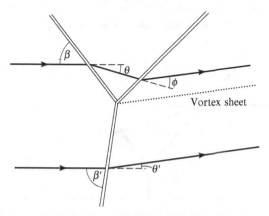

Fig. 6.13 Shock interaction; case 2.

possible for a weak shock, when $M_1 \sin \beta_1 \downarrow 1$, and in this case, the angle of reflection $\beta_3 \to \beta_1$ and the reflection is specular as in Figure 4.3 (see Exercise 6.20). However, Figure 6.7 shows that there is a maximum angle θ_{max} through which a flow with Mach number M_1 can be turned by a single shock and θ_{max} decreases as M_1 decreases from ∞. Inevitably, the Mach number M_2 behind the shock will be less than M_1, and so there will be a range of values of (M_1, β_1) for which θ will exceed the maximum deflection angle corresponding to M_2. Experimental observations [53] show that it is this case that leads to the phenomenon of Mach reflection. Since the shock wave below the triple point, which is called the *Mach stem*, is observed to be nearly normal to the wall, the flow behind it is subsonic and thus influenced by the conditions downstream; this means that the downstream shock and vortex sheet shown in Figure 6.15 can only be straight locally near the triple point and the streamlines in the subsonic region will bend to allow the flow to eventually become parallel to the wall. (This is a similar effect to that noted in the footnote in Section 6.1.3.)

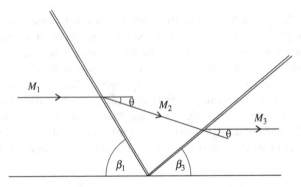

Fig. 6.14 Shock reflection at a wall: case I

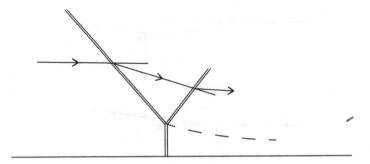

Fig. 6.15 Shock reflection at a wall: case II

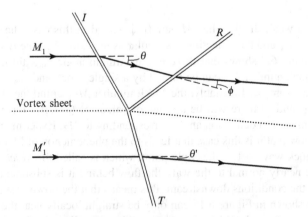

Fig. 6.16 Shock reflection and transmission from a vortex sheet.

Finally, we mention the possibility of achieving shock attenuation as a result of impact with a vortex sheet, as shown in Figure 6.16. As detailed in Exercise 6.21, the strength of the transmitted shock T is less than that of the incident shock I, and this has been proposed as a method of ameliorating sonic boom. In this context, we note that annoying oblique "bow shocks" are inevitably generated by supersonic aircraft with the one exception of the *Busemann Biplane*, which generates a finite system of shocks as shown in Figure 6.17 and is described in Liepmann and Roshko [36]. Alas, this biplane is only of academic interest since it generates no lift!

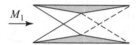

Fig. 6.17 Shock waves generated by the Busemann Biplane.

6.2.3 Steady Quasi-one-dimensional Gas Flow

In Section 4.7.2, we showed how the equations of gas dynamics can be reduced to three algebraic equations (4.60)–(4.62) in the case of flow in a nozzle of slowly varying cross-section $A(x)$. Such flows are called *quasi-one-dimensional*, and our knowledge of shock waves now enables us to understand the flow through a converging-diverging nozzle more fully.

We first use equations (4.60)–(4.62) to write A and the pressure p in terms of the Mach number $M = u/c$ as

$$A = \frac{m}{\rho_0 c_0 M}\left\{1 + \frac{\gamma - 1}{2}M^2\right\}^{(\gamma+1)/(2(\gamma-1))} \tag{6.56}$$

and

$$p = p_0\left\{1 + \frac{\gamma - 1}{2}M^2\right\}^{-\gamma/(\gamma-1)}. \tag{6.57}$$

Here p_0, ρ_0 and c_0 are the pressure, density and speed of sound in a large reservoir which feeds the nozzle. Plotting A against M as in Figure 6.18, we see that the minimum value A_c of this function can only be attained when M is unity and the flow

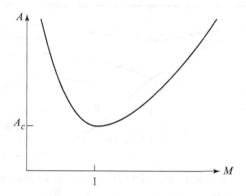

Fig. 6.18 Variation of A with M from (6.56).

is sonic.[6] Hence, we are led to consider the flow from a high pressure reservoir into a converging-diverging nozzle, known as a *Laval nozzle*, as shown in Figure 6.19.

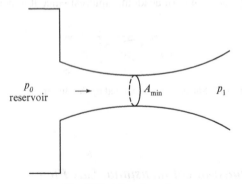

Fig. 6.19 A Laval nozzle.

If the minimum cross-sectional area of the nozzle, A_{min}, is greater than A_c, then the flow will always be subsonic, whereas if we can arrange for A_{min} to equal A_c, it might be possible for the flow to become supersonic in the divergent part of the nozzle. If this can be done, it leads to a design for a supersonic wind tunnel. The flow rate m will be controlled by the downstream pressure p_1 imposed at the end of the nozzle and, by decreasing p_1 from p_0, we can arrive at the exit pressure p_c for which $A_{min} = A_c$. The question then arises as to "what happens when $p_1 < p_c$?"

In Figure 6.20, we show how the pressure along the nozzle varies for different values of p_1. We can see that if p_1 is between p_0 and p_c, then the flow will remain

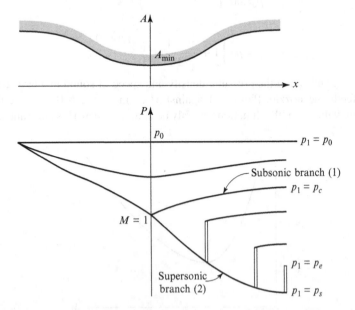

Fig. 6.20 Pressure variation in a nozzle for varying values of the downstream pressure p_1.

---

[6] Note that we have implicitly assumed that the flow is compressible. Clearly an incompressible flow can flow smoothly through any slowly varying nozzle.

subsonic throughout. When $p_1 = p_c$, the Mach number will be unity at the throat, although the flow will revert to being subsonic in $x > 0$ as indicated by branch (1) in Figure 6.20. However, the key observation is that there is *another* possible solution with $M = 1$ at the throat; this will correspond to branch (2) with supersonic flow in $x > 0$ and it occurs when $p_1 = p_s < p_c$; the values of p_c and p_s can be found from (6.56) and (6.57). When $p_1 < p_c$, the flow is said to be *choked*, and the new phenomenon in such flows is the presence of a shock wave in the supersonic flow downstream of the throat; the flow upstream of the throat is the same for all $p_1 < p_c$. As p_1 decreases from p_c, this shock moves to the right and eventually reaches the end of the nozzle when p_1 attains yet another critical value p_e. For values of p_1 between p_s and p_e, the shock wave is ejected into the downstream atmosphere in a complicated three-dimensional flow involving multiple shock waves. Furthermore, if $p_1 < p_s$, the flow downstream will contain a series of Prandtl–Meyer expansion fans as discussed in Chapman [7].

6.2.4 *Shock Waves with Chemical Reactions

The violence inflicted on gas particles as they pass through a shock wave can frequently induce chemical reactions, the most awesome of which is when a combustible gas undergoes an intense exothermic reaction as it encounters the temperature, pressure and density rise at the shock. Such a configuration is called a *detonation*, and here we mention the simplest model for such detonations. We simply sweep all the chemistry aside and assert that, while the mass and momentum conservation laws still apply at the shock, energy is gained at a prescribed rate E per unit mass. Thus, (6.23) becomes

$$\left[\frac{\gamma p}{(\gamma - 1)\rho} + \frac{1}{2}(\dot{X}_s - u)^2 \right] = E. \tag{6.58}$$

This makes our mathematical analysis, which is complicated enough when $E = 0$, even more difficult to present lucidly. However, great insight can be obtained by noticing that, from (6.24) and (6.25), when $E = 0$, the ratios $\tilde{p} = p_2/p_1$, $\tilde{\rho} = \rho_2/\rho_1$, satisfy

$$\tilde{p} = \frac{(\gamma + 1) - (\gamma - 1)/\tilde{\rho}}{(\gamma + 1)/\tilde{\rho} - (\gamma - 1)}, \tag{6.59}$$

where suffix 1 is upstream in the unreacted gas. Thus, the point $(\tilde{p}, \tilde{\rho}^{-1})$ lies on a hyperbola in the $(\tilde{p}, \tilde{\rho}^{-1})$ plane called the *Chapman–Jouguet* curve as shown in Figure 6.21. The point (1,1) corresponds to the upstream condition, and the unique compressive solution satisfying the physically acceptable Rankine–Hugoniot condition lies on the part of the hyperbola indicated by the solid line.

Now let us reintroduce E. Equation (6.59) becomes

$$\tilde{p} = \frac{\gamma + 1 - (\gamma - 1)/\tilde{\rho} + 2\gamma(\gamma - 1)E/c_1^2}{(\gamma + 1)/\tilde{\rho} - (\gamma - 1)}, \tag{6.60}$$

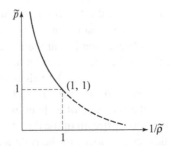

Fig. 6.21 The Chapman–Jouguet curve.

so that the Chapman–Jouguet curve is displaced upwards, as shown in Figure 6.22. The point $(1,1)$ still represents the upstream condition, but our selection criteria that led us to reject the dashed segment of the hyperbola in Figure 6.22 are no longer available; even the simplest entropy argument would be difficult in the presence of the reaction. It is easy to see that the dashed segment where $\tilde{p} > 1$ and $\tilde{\rho} < 1$ cannot satisfy the Rankine–Hugoniot conditions, but the lower branch where $\tilde{p} < 1$ and $\tilde{\rho} < 1$ cannot be ruled out. This branch represents a *deflagration* and it can be observed in certain circumstances. Further details about this theory can be found in Courant and Friedrichs [13].

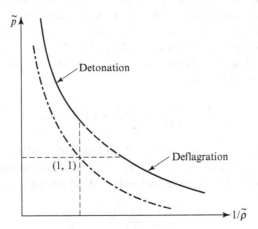

Fig. 6.22 The Chapman–Jouguet curve with constant energy addition.

6.2.5 Open Channel Flow

Our last example concerns the steady flow of shallow water along a straight channel with a horizontal bottom but having slowly varying width. This problem is very similar to the gas flow studied in Section 6.2.3 if we make the usual shallow water assumptions that pressure is hydrostatic and that the flow is quasi-one-dimensional. If the width of the channel is $b(x)$, the depth of the water is η and the horizontal velocity is u, conservation of mass gives

$$ub\eta = q, \tag{6.61}$$

where q is the constant flux along the channel. In addition, Bernoulli's equation on the surface streamline leads to

$$\frac{1}{2}u^2 + g\eta = gH, \tag{6.62}$$

where H is the pressure head, which is the depth of the water in a large reservoir which feeds the channel. We write these equations in terms of the Froude number

$$F = \frac{u}{\sqrt{g\eta}}, \tag{6.63}$$

to get

$$\eta = \frac{H}{1 + \frac{1}{2}F^2},$$

$$u = F\left(\frac{gH}{1 + \frac{1}{2}F^2}\right)^{1/2}$$

and

$$b = \frac{q}{g^{1/2}H^{3/2}F}\left(1 + \frac{1}{2}F^2\right)^{3/2}.$$

Plotting b as a function of F in Figure 6.23, shows that b has a minimum when $F = 1$, in the same way that A in (6.56) is minimised when $M = 1$.

As for the quasi-one-dimensional gas flow, it is interesting to consider flow in a channel whose width first converges and then diverges. From Figure 6.23, we see that it is possible to have a flow that changes from subcritical ($F < 1$) to supercritical ($F > 1$) as long as $F = 1$ at the point of minimum width and, in that case, the depth will be a decreasing function of distance along the channel. As in Section 6.2.3, it may be necessary to introduce a steady hydraulic jump into the supercritical flow in the divergent part of the channel in order to satisfy the downstream conditions.

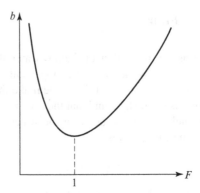

Fig. 6.23 Variation of b with F.

6.3 *Further Limitations of Linearised Gas Dynamics

In this section, we first revisit the linear theory for thin wings in transonic and supersonic two-dimensional steady flow that were derived in Section 4.7.3. We will show how we can (i) discuss the transonic case when $M \simeq 1$ and (ii) extend the supersonic theory to the far field. Finally, we return to study the flow past a thin wing when $M \gg 1$.

6.3.1 Transonic Flow

In Section 4.7.3, we have already anticipated trouble with our thin-wing theories when the free stream Mach number M is close to unity. To derive a small disturbance theory that is valid in the transonic regime, we have to return to (5.17). Writing the velocity potential as $Ux + \phi$, (5.17) becomes

$$\left(c^2 - \left(U + \frac{\partial \phi}{\partial x}\right)^2\right)\frac{\partial^2 \phi}{\partial x^2} - 2\left(U + \frac{\partial \phi}{\partial x}\right)\frac{\partial \phi}{\partial y} \cdot \frac{\partial^2 \phi}{\partial x \partial y} + \left(c^2 - \left(\frac{\partial \phi}{\partial y}\right)^2\right)\frac{\partial^2 \phi}{\partial y^2} = 0,$$

where

$$c^2 = c_0^2 + \frac{\gamma - 1}{2}U^2 - \frac{(\gamma - 1)}{2}\left(\left(U + \frac{\partial \phi}{\partial x}\right)^2 + \left(\frac{\partial \phi}{\partial y}\right)^2\right).$$

We can still assume that ϕ and its derivatives are small, but we must remember that $1 - U^2/c_0^2$ is also small now. Thus, the coefficient of $\partial^2 \phi/\partial x^2$ is

$$c^2 - \left(U + \frac{\partial \phi}{\partial x}\right)^2 \sim c_0^2 - U^2 - (\gamma + 1)U\frac{\partial \phi}{\partial x},$$

and we are led to the following equation for small disturbances in transonic flow:

$$(1 - M^2)\frac{\partial^2 \phi}{\partial x^2} + \frac{\partial^2 \phi}{\partial y^2} = \frac{(\gamma + 1)M}{c_0}\frac{\partial \phi}{\partial x}\frac{\partial^2 \phi}{\partial x^2}. \tag{6.64}$$

This equation is nonlinear, and its qualitative properties are much less well understood than are the corresponding subsonic and supersonic equations. The basic difficulty is that when $M = 1$, (6.64) changes from being hyperbolic to elliptic when $\partial \phi / \partial x$ changes sign. This equation is said to be of *mixed type* and some idea of the possible behaviour of solutions of such equations can be gained by considering the linear mixed-type Tricomi equation as in Exercise 4.1 (see Garabedian [20]).

We should also mention that we must be prepared for solutions of (6.64) to contain shocks. Indeed, the numerical solution of slightly subsonic flow past a thin wing shows both a supersonic region and a shock as illustrated in Figure 6.24. Fortunately, these shocks are weak enough for any entropy jump across them to be neglected and hence for the assumption of irrotationality to be justified.

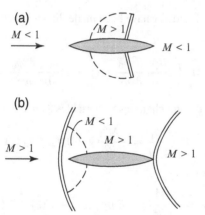

Fig. 6.24 Transonic flow past a thin wing. (a) Lower transonic regime, (b) Upper transonic regime

More importantly, (6.64) forms the basis for correcting formulas such as (4.68), (4.70) and (4.71) which become invalid as B or $\beta \to 0$ (see [24] for further details). The equation also models the local flow near the sonic line in situations such as those illustrated in Figure 6.9.

Although all transonic phenomena ultimately require a nonlinear model close to points where the Mach number is unity, we have already seen, in Section 4.9, that linear theory can be used to identify such points. Another example is when a source moves through an atmosphere in which the sound speed varies so that it is less than the speed of the source near to the source but greater than it further away. Then, there will be a sonic boundary in the flow near which a linear Tricomi model such as that proposed in Exercise 4.1 will apply.

6.3.2 The Far Field for Flow past a Thin Wing

The linearised theory of Section 4.7.3 showed that, in supersonic flow past a thin wing, the disturbance is confined between characteristics emanating from the leading and trailing edges of the body and that this inevitably entailed some sort of discontinuity across these characteristics. A calculation can be performed to show that weak shocks or expansion fans will in fact lie near these characteristics. Usually, such discontinuities may be neglected when calculating the aerodynamic forces. But, from an environmental point of view, the effect of these discontinuities needs to be understood in the "far field" of the wing in order to assess noise at ground level due to supersonic aircraft.

We first need to assess the region of validity of the linearised solution (4.69)–(4.70), and so we go to the next term in the expansion for the velocity potential by writing $\phi \sim Ux + lU(\varepsilon\phi_0 + \varepsilon^2\phi_1 + \cdots)$ in equation (5.17). Then, ϕ_0 will be given by (4.70) so that, above the wing,

$$\phi_0 = -B^{-1}f_+(x - By) \text{ for } y > 0,$$

where $B^2 = U^2/c_0^2 - 1$ and x, y have been made dimensionless with l. The equation for ϕ_1 is

$$B^2\frac{\partial^2\phi_1}{\partial x^2} - \frac{\partial^2\phi_1}{\partial y^2} = -M^2\left(2\frac{\partial\phi_0}{\partial y}\frac{\partial^2\phi_0}{\partial x\partial y} + (\gamma+1)\frac{\partial\phi_0}{\partial x}\frac{\partial^2\phi_0}{\partial x^2} + (\gamma-1)\frac{\partial\phi_0}{\partial x}\frac{\partial^2\phi_0}{\partial y^2}\right).$$

Now, substituting for ϕ_0 and changing to variables $\xi = x - By$, $\eta = x + By$ leads to

$$4B^2\frac{\partial^2\phi_1}{\partial\xi\partial\eta} = -\frac{M^4}{B^2}(\gamma+1)f'_+(\xi)f''_+(\xi)$$

in $y > 0$, and the solution is

$$\phi_1 = -\frac{M^4}{8B^4}(\gamma+1)\eta f'_+(\xi)^2 + F(\xi) + G(\eta),$$

where F and G are arbitrary functions. Thus, as x, y increase along the lines where ξ is constant, $\varepsilon^2\phi_1$ will inevitably be comparable in size with $\varepsilon\phi_0$ when x and y are $O(\varepsilon^{-1})$. Physically, what is happening is that nonlinearity is, inevitably, modulating the linear theory, as we have already seen in Section 5.3 and in Exercise 5.11, for example. The manifestation of this modulation is that, to second order in ε, the leading shock is neither straight nor is it a characteristic; instead, it is slightly curved and is intersected by the characteristics as sketched in Figure 6.25. This phenomenon first becomes apparent at distances of $O(\varepsilon^{-1}l)$ from the wing as we will now see (further details are given in Van Dyke [52]).

To get a quantitative description of the far field flow, we can use the same method as that of Section 5.3.1. Guided by the analysis above, we change to variables $\xi = x - By$, $Y = \varepsilon y$ in (5.17) before writing $\phi \sim Ux + lU\varepsilon\Phi$, which leads directly to

$$2B\frac{\partial^2\Phi}{\partial\xi\partial Y} = -(\gamma+1)M^4\frac{\partial\Phi}{\partial\xi}\frac{\partial^2\Phi}{\partial\xi^2} + O(\varepsilon).$$

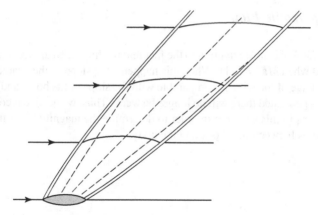

Fig. 6.25 The far field flow past a thin wing in supersonic flow.

Thus, the perturbation velocity $u = \partial \Phi / \partial \xi$ satisfies the now-familiar kinematic wave equation

$$\frac{\partial u}{\partial Y} + \frac{(\gamma + 1)M^4}{2B} u \frac{\partial u}{\partial \xi} = 0 \qquad (6.65)$$

with the initial condition $u = -(1/B)f'(\xi)$ on $Y = 0$. We can even write down the explicit solution (shock waves and all!) in the case where the wing profile is parabolic and at zero incidence (Exercise 6.23). In this case, both the leading and trailing characteristics are weak *shocks*[7] and this solution reveals the famous "*N*-wave" solution, which can be shown to be the "generic" solution as $Y \to \infty$, and gives a pressure profile as shown in Figure 6.26. The shocks weaken as $O(Y^{-1/2})$ at the same time as the expansion wave spreads parabolically in ξ. This explains the "double bang" that is sometimes heard on the ground when an aircraft flies supersonically at altitude, but possibly many kilometres away horizontally.

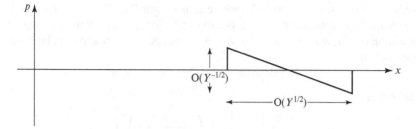

Fig. 6.26 The pressure in the far field.

---

[7] Depending on the aerofoil shape and angle of incidence, the leading edge could emit either two weak shock waves or a weak shock and a weak expansion wave.

6.3.3 Hypersonic Flow

In Section 4.7.3, we drew attention to the fact that the linearised approximation is no longer valid when $M\varepsilon = O(1)$. When this happens, the slope of the characteristic at the leading edge of the wing is comparable with the slope of the body, and hence the shock that is generated there will no longer be weak. Thus, we need to reconsider the shock relations in this situation in order to determine the magnitudes of the various flow quantities between the shock and the body.

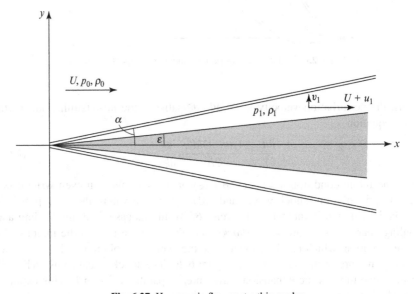

Fig. 6.27 Hypersonic flow past a thin wedge.

Suppose a two-dimensional steady shock is generated by flow past a symmetric wedge as illustrated in Figure 6.27, where 2ε is the small angle of the wedge and α, the angle of inclination of the shock, is also small. Now, from the Rankine–Hugoniot condition (6.46), with $\beta = \alpha$ and $\theta = \varepsilon$, the assumption that α and ε are both small implies that

$$(\gamma + 1)M^2\alpha(\alpha - \varepsilon) = 2 + (\gamma - 1)M^2\varepsilon^2,$$

and, hence,

$$\alpha = \frac{\gamma + 1}{4}\varepsilon\left(1 + \left[1 + \left(\frac{4}{(\gamma + 1)\varepsilon M}\right)^2\right]^{1/2}\right).$$

Thus, α is of $O(\varepsilon)$ as long as $M\varepsilon \geq O(1)$. On the body, the boundary condition is

$$v_1 = (U + u_1)\varepsilon,$$

and, parallel to the shock, the condition (6.36) gives

$$u_1 \cos \alpha + v_1 \sin \alpha = 0,$$

so that we can deduce that $v_1/U = O(\varepsilon)$ and $u_1/U = O(\varepsilon^2)$. Furthermore, from the Rankine–Hugoniot equations (6.43) and (6.44), it is easy to show that

$$\frac{p_1}{p_0} = O(1) \quad \text{and} \quad \frac{\rho_1}{\rho_0} = O(1).$$

Using these estimates and remembering that $p_0 = O(\rho_0 u^2 \varepsilon^2)$, we can now scale the variables in the flow between the shock and a more general body whose slope is of $O(\varepsilon)$ by writing

$$u = U(1 + \varepsilon^2 \bar{u}), \quad v = U\varepsilon \bar{v}, \quad p = \rho_0 U^2 \varepsilon^2 \bar{p}, \quad \rho = \rho_0 \bar{\rho} \quad \text{and} \quad y = \varepsilon l \bar{y}, \quad x = l \bar{x},$$

where l is the streamwise length scale of the body. Then, we can expect that the barred variables will all be $O(1)$ behind the shock and the first approximation to the equations (2.6), (2.7) and (2.24) will be

$$\frac{\partial \bar{\rho}}{\partial \bar{x}} + \frac{\partial}{\partial \bar{y}}(\bar{\rho}\bar{v}) = 0, \tag{6.66}$$

$$\frac{\partial \bar{u}}{\partial \bar{x}} + \bar{v}\frac{\partial \bar{u}}{\partial \bar{y}} = -\frac{1}{\bar{\rho}}\frac{\partial \bar{p}}{\partial \bar{x}}, \tag{6.67}$$

$$\frac{\partial \bar{v}}{\partial \bar{x}} + \bar{v}\frac{\partial \bar{v}}{\partial \bar{y}} = -\frac{1}{\bar{\rho}}\frac{\partial \bar{p}}{\partial \bar{y}}, \tag{6.68}$$

and

$$\left(\frac{\partial}{\partial \bar{x}} + \bar{v}\frac{\partial}{\partial \bar{y}}\right)\left(\frac{\bar{p}}{\bar{\rho}^\gamma}\right) = 0. \tag{6.69}$$

Equations (6.66), (6.68) and (6.69) do not depend on $\bar{u}$, even though they model a two-dimensional flow, and hence they are exactly equivalent, with a suitable change of notation, to the *one-dimensional* unsteady equations (5.3)–(5.5). Moreover, the approximate boundary condition for a body given by $\bar{y} = f(\bar{x})$ is $\bar{v} = f'(\bar{x})$ on $\bar{y} = f(\bar{x})$, and this also translates into the boundary condition for a piston moving with velocity $f'(t)$. Better yet, Exercise 6.24 reveals that shock conditions turn out to be such that the problem for $\bar{p}$, $\bar{\rho}$ and $\bar{v}$ is mathematically *identical* to that of the unsteady flow caused by such a piston. This analogy is called *hypersonic similitude*, and a similar analogue exists between the steady three-dimensional hypersonic flow past an axisymmetric slender body and a two-dimensional unsteady gas dynamics problem. Thus, we have found a class of gas dynamic flows in which time and space can be identified with each other.

If $M\varepsilon \gg 1$, then the shock is strong enough for the shock relations to simplify still further. Suppose, in particular, that we consider such a flow past a two-dimensional power-law body given by $\bar{y} = C\bar{x}^k$. Thus, as shown in Exercise 6.25, the shock relations are simple enough for us to find a similarity solution which depends only on the variable $\bar{y}/\bar{x}^k$. The principle of hypersonic similitude now states that the problem of a strong shock driven by a piston whose position is proportional to t^k also reduces to ordinary differential equations in the single variable x/t^k; happily, a similar reduction to ordinary differential equations also occurs for radially or spherically symmetric pistons whose position is proportional to t^k.

A dramatic example of such an unsteady flow which can be solved explicitly is that of the *blast wave* caused by a sudden explosion at $t = 0$ at the origin. Such an explosion results in the sudden release of a large amount of energy E and causes a strong shock to expand into the initially quiescent surrounding gas. If the shock position is given by $r = R(t)$ and the density, pressure and velocity of the gas just behind the shock are ρ_1, p_1 and v_1, then, since the shock is strong, the Rankine–Hugoniot equations (6.24)–(6.26) reduce to

$$\rho_1 = \rho_0 \frac{(\gamma + 1)}{(\gamma - 1)},$$

$$v_1 = \frac{2\dot{R}}{\gamma + 1}$$

and

$$p_1 = \frac{2\rho_0 \dot{R}^2}{\gamma + 1},$$

where ρ_0 is the ambient density. Thus, the only physical quantities that enter this problem are r, t, E and ρ_0. In the three-dimensional case, the only nondimensional variable that can be constructed from these quantities is

$$\xi = r\left(\frac{\rho_0}{Et^2}\right)^{1/5}.$$

Hence, ρ/ρ_0 will be a function of ξ and, since ρ/ρ_0 is constant at the shock, we retrieve the famous result that an atomic bomb blast grows such that $R/t^{2/5} = K = $ constant. In order to find K, we must solve the equations of unsteady gas flow behind the shock, which we know reduce to ordinary differential equations in ξ with boundary conditions given on the shock. One integral of the motion can be found immediately by observing that the total energy behind the shock remains constant and equal to E, and this leads us to a relation between E and K. Thus, one can estimate the energy release in an atomic explosion by observing the growth of the emitted blast wave. This theory also reveals that such blast waves could be generated by a spherical piston expanding so that the radius grows like $t^{2/5}$.

Going down a dimension, we can consider a cylindrical blast wave caused by a line explosion releasing a fixed amount of energy per unit length. Then the dimensional argument shows that the blast wave expands so that the radius is proportional to $t^{1/2}$. Using hypersonic similitude, we can deduce that the same solution holds for the hypersonic flow past a slender axisymmetric body whose radius is proportional to $x^{1/2}$. There is one caveat: this body has a blunt nose and the flow near the nose will not be susceptible to hypersonic small disturbance theory. An asymptotic analysis reveals that there is a thin *entropy layer* near the body where a new solution is needed, but luckily this does not affect the position of the shock to first order. For this case, it is also possible to show that the energy release in the cylindrical blast wave is directly related to the drag on the nose of the blunt body in hypersonic flow (Anderson [2]).

On a less warlike note, we remark that a similar kind of similitude applies to high-speed ocean transport. As shown in Exercise 6.26, the waves caused by a high-speed (high Froude number) slender ship moving with constant velocity are equivalent to the unsteady waves produced by a wavemaker at the end of a one-dimensional water tank. Whereas hypersonic similitude demands that $M\varepsilon = O(1)$, it transpires that slender ship theory is valid when $F\varepsilon^2 = O(1)$, where F is the Froude number and ε the slenderness parameter.

Finally, when we recall the analogy between gas dynamics and shallow water theory, we might ask if there is any link between hypersonic gas dynamics and the high-Froude-number limit that led to the theory of delta shocks (Section 6.1.5). In hypersonic flow, Figure 6.27 reveals the existence of a thin high-density region between the shock and body but, instead of tending to a delta shock, this region maintains a finite thickness as $M \to \infty$. Hence, it cannot be modelled as a delta shock except in the double limit $M \to \infty$, $\gamma \to 1$.[8] This limit leads to the *Newtonian Theory of Hypersonic Flow* in which there is an infinitely thin "shock layer" on the body, in which the pressure distribution is precisely that found by Newton [40] in his particle-impact model for hydrodynamic drag. For more information about hypersonic flow see [25].

6.4 *Shocks in Solids and Plasmas

We have claimed in the introduction to this book that fluid dynamics is the best vehicle for learning physical applied mathematical methodologies, and we end this chapter by describing how the nonlinear theories presented in Chapters 5 and 6 can be applied to some of the other models described in Chapters 3 and 4.

---

[8] This Newtonian limit is not so unrealistic because molecular effects in hypersonic flows cause γ to decrease below 1.4 for air.

6.4.1 The Piston Problem in Elastoplasticity

We now consider a simple one-dimensional problem for an elastoplastic material using the equations derived in Section 3.8.4. We suppose that a prescribed compressive pressure $P_1(t)$ is imposed at $a = 0$ on a semi-infinite block of material which is at rest for $t < 0$. As the wavefront caused by the impact at $t = 0$ moves into the material, it will first be deformed elastically, possibly generating a shock wave, and then, if P_1 is sufficiently large, the yield criterion will be attained. In order to simplify the problem, we will consider the case where $P_1 \gg \sigma_Y$ and let $\sigma_Y \to 0$ in (3.81). This means that the material becomes elastoplastic almost as soon as the motion starts and we use equations (3.80), (3.87) and the "equation of state" (3.92) to describe this plastic flow. We can rewrite these equations in terms of $u = \partial x / \partial t$ and $\hat{\rho} = \rho / \rho_0$ to get

$$\frac{\partial \hat{\rho}}{\partial t} + \hat{\rho}^2 \frac{\partial u}{\partial a} = 0, \tag{6.70}$$

$$\rho_0 \frac{\partial u}{\partial t} = \frac{\partial \tau_1}{\partial a} \tag{6.71}$$

and

$$\tau_1 = -(3\lambda + 2\mu)\hat{\rho}^{\frac{1}{3}}(\hat{\rho}^{\frac{1}{3}} - 1). \tag{6.72}$$

As we noted earlier, if we write $\tau_1 = -p$, these equations are exactly those of a barotropic gas written in Lagrangian coordinates.

The initial conditions are

$$u = 0, \quad \hat{\rho} = 1 \text{ at } t = 0 \text{ for } a > 0,$$

and the boundary condition is

$$\tau_1 = -P_1(t) \text{ at } a = 0 \text{ for } t > 0.$$

These equations only hold in the plastic region of the (a, t) plane and this region will be separated from the initially undisturbed region by a "free boundary". This is *not* the same as the discontinuities that we have met earlier in this chapter since, although the equations of mass and momentum, (6.70) and (6.71), hold on both sides of the free boundary, ahead of the free boundary the equation of state (6.72) will be replaced by the constitutive laws of elasticity. Nevertheless, we can apply the ideas of Section 6.1.1 to equations (6.70) and (6.71) to derive appropriate "jump relations", namely,

$$\dot{A} = \frac{-[u]}{[1/\hat{\rho}]} = -\frac{[\tau_1]}{[\rho_0 u]},$$

where the free boundary is given by $a = A(t)$. The material ahead of the free boundary is at rest with $u = 0$ and $\hat{\rho} = 1$, so that

$$\dot{A} = \frac{p_b}{\rho_0 u_b} = \frac{-u_b}{1/\hat{\rho}_b - 1}, \tag{6.73}$$

where the subscript "b" denotes the value of a variable downstream of $a = A(t)$. In the simple case when the applied pressure P_1 is a constant impulsive stress, the flow behind the discontinuity will be uniform with $p_b = P_1$ and, by solving equations (6.73) and (6.72), u_b, $\hat{\rho}_b$ and $\dot{A}$ can be determined. The implications of this model are discussed in detail in [29] where it is shown that the response is similar to that of a barotropic gas until the free boundary overtakes the leading elastic wave. Such "overdriven" compressions are not well understood.

This example also raises the interesting question of the relation between the Rankine–Hugoniot equations in Lagrangian variables rather than the Eulerian variables we have used for gas dynamics. It can be shown (Exercise 6.27) that, if the free boundary is given by $x = X_f(t)$, then

$$\rho_b(\dot{X}_f - u_b) = \rho_0\dot{A},$$

and it is then straightforward to see that the jump relations (6.73) are exactly equivalent to (6.16) and (6.18). Note that, because the plastic flow is analogous to *homentropic* barotropic flow, a shock in a plastic flow is more like a bore in shallow water than a shock in gas dynamics which will always involve a jump in entropy.

This research theme is covered in much more detail in Davison [15] and Germain and Lee [21].

6.4.2 Shock Tube Problems in Cold Plasmas

The fact that the plasma model described in Section 3.8.2 involves Poisson's equation (3.50), which is elliptic and therefore needs to be solved globally, makes it easier to study shock tube problems rather than piston problems for plasmas (and they are also easier to study experimentally). We therefore consider the configuration illustrated in Figure 6.10, where the tube is initially filled with a plasma and we adopt the cold ion model governed by (3.54), (3.56) and (3.57). We suppose that the ion gas is initially in equilibrium with

$$n_i(x,0) = \begin{cases} 1, & \text{for } x < 0 \\ n_r, & \text{for } x > 0 \end{cases},$$

where $n_r < 1$, and, at time $t = 0$, the diaphragm at $x = 0$ is ruptured.

Analytical progress is only possible if we assume that n_r is close to unity and, by linearising as in (3.58)–(3.60) with $\tilde{n} = (1 - n_r)\hat{n}$, we obtain

$$\frac{\partial^4\hat{n}}{\partial x^2 \partial t^2} - \frac{\partial^2\hat{n}}{\partial t^2} + \frac{\partial^2\hat{n}}{\partial x^2} = 0 \tag{6.74}$$

with

$$\hat{n}(x,0) = \begin{cases} 0, & x < 0 \\ -1, & x > 0 \end{cases}. \tag{6.75}$$

The relevant dispersion relation is (4.33) and, using Fourier transforms, we find (Exercise 6.28) that

$$\hat{n} = -\frac{1}{2} - \frac{1}{2\pi} \int_0^\infty \left[\sin\left(kx + \frac{kt}{\sqrt{1+k^2}} \right) + \sin\left(kx - \frac{kt}{\sqrt{1+k^2}} \right) \right] \frac{dk}{k}, \quad (6.76)$$

which is illustrated in Figure 6.28. The oscillations observed arise from dispersion and are in strong contrast to the monotonic pressure variation found in the solution to the gas dynamic shock tube problem in Section 6.2.1. Numerical solutions of the nonlinear equations (3.54) and (3.56), (3.57) show that these oscillations persist as n_r becomes smaller but that they tend to focus near the leading wavefront $x = t$, as shown in Figure 6.29.[9]

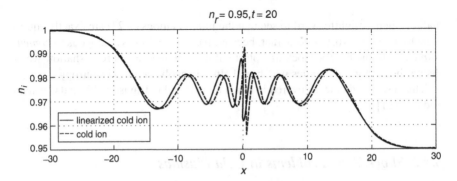

Fig. 6.28 Plot of n_i from (6.78).

We conclude by mentioning the surprising phenomenon that can emerge as a result of the random fluctuations that occur when the ions have non-zero temperature. Now we have to revert to the full Vlasov–Poisson system (3.53), (3.54) and (3.55) with the initial condition for the distribution function f_i given by the Maxwellian (3.71), which, in dimensionless variables, can be written as

$$f_i(x, 0, u) = \frac{1}{\sqrt{\pi \varepsilon}} e^{-u^2/\varepsilon},$$

where ε represents the ion temperature and is small. The numerical solution of this system closely resembles the cold ion solutions in Figure 6.29 when n_r is not too small, but for $n_r \sim O(10^{-1})$, there is an increasingly large deviation from these results, especially in u, near the leading edge of the wave.

To understand this new phenomenon, we need to visualise the distribution function $f_i(x, t, u)$, and we do this by plotting the level curves of f_i in x, u for a fixed time; these almost all lie in the regions between the bold curves in Figure 6.30. The largest values of f_i are around the central curves of the shaded region, and we see

---

[9] Figures 6.28, 6.29 and 6.30 are taken from [47], see also [38].

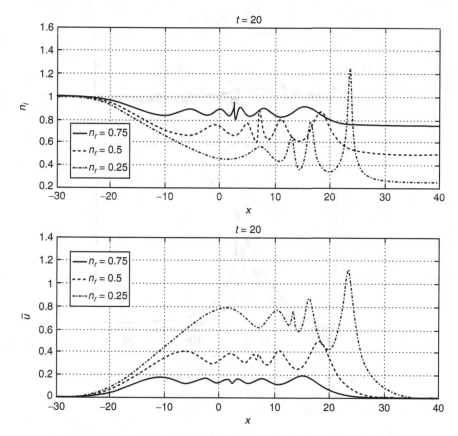

Fig. 6.29 Numerical solution for n_i and $\bar{u}$ in a shock tube when n_r is $O(1)$.

that when n_r is close to 1, f_i is only non-zero near $u = 0$ and thus remains approximately Maxwellian. However, for small n_r, f_i is not only not Maxwellian but it also becomes *multivalued* for a range of x. This suggests that the ions have split into at least two distinct populations, one with much higher mean velocity than the other. Indeed, the curves in Figure 6.30 are reminiscent of a shallow water wave breaking on a beach, and we may speculate that our cold ion model breaks down for small n_r in the same way that shallow water theory breaks down on a beach; both situations need to be resolved using either multivalued functions or, better still, a Lagrangian model. This is yet another example of the way in which the mathematical concepts that have been expounded in this book can shed new light on wave phenomena across the physical sciences.

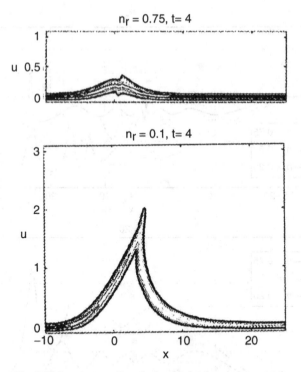

Fig. 6.30 Level curves for u in the shock tube when $n_r \ll 1$

Exercises

R 6.1 Derive (6.21) from (6.16) and deduce that, for any function f,

$$[f\rho u] - \dot{X}_s[f\rho] = [f]m,$$

where $m = \rho(u - \dot{X}_s)$. Then put $f = \dot{X}_s$ to derive (6.22) from (6.18). Finally put $f = p/\rho$ to show that (6.20) implies that

$$\frac{m\gamma}{\gamma - 1}\left[\frac{p}{\rho}\right] + \left[\frac{1}{2}\rho u^3\right] + \dot{X}_s[p] = \dot{X}_s\left[\frac{1}{2}\rho u^2\right].$$

Using (6.22) for $[p]$ and writing $\left[\frac{1}{2}\rho u^2(u - \dot{X}_s)\right] = m\left[\frac{1}{2}u^2\right]$, deduce (6.23).

R 6.2 Writing $M_i = (\dot{X}_s - u_i)/c_i$, show that (6.21) implies that $[\rho^2 M^2 c^2] = 0$, where $c^2 = \gamma p/\rho$. Deduce that $p_2\rho_2/p_1\rho_1 = M_1^2/M_2^2$.
Show that (6.22) implies that $[p(1 + \gamma M^2)] = 0$ and hence that

$$\frac{p_2}{p_1} = \frac{1 + \gamma M_1^2}{1 + \gamma M_2^2}.$$

Show that (6.23) implies that

$$\left[\frac{p}{\rho}(1 + \frac{\gamma - 1}{2}M^2)\right] = 0$$

and hence that

$$\frac{p_2\rho_1}{p_1\rho_2} = \frac{2 + (\gamma - 1)M_1^2}{2 + (\gamma - 1)M_2^2}.$$

Combine these results to show that

$$(M_2^2 - M_1^2)[M_2^2(2\gamma M_1^2 - (\gamma - 1)) - ((\gamma - 1)M_1^2 + 2)] = 0,$$

and hence derive (6.24)–(6.26).

6.3 Show that the entropy jump across a shock is

$$[S] = c_V\left[\log p/\rho^\gamma\right].$$

Use (6.24)–(6.26) to write this as

$$[S] = c_V\left(\log(2\gamma x - (\gamma - 1)) + \gamma\log\left(\frac{2}{x} + \gamma - 1\right) - (\gamma + 1)\log(\gamma + 1)\right),$$

where $x = M_1^2$. Show further that

$$\frac{1}{c_V}\frac{d[S]}{dx} = 2\gamma(\gamma - 1)\frac{(x - 1)^2}{(2\gamma x - (\gamma - 1))(2/x + \gamma - 1)x^2}.$$

Deduce that

(i) $[S] > 0$ when $x > 1$,
(ii) $[S] = O((x - 1)^3)$ when $x \downarrow 1$.

6.4 From (6.21) and (6.22) show that across a normal shock wave

$$(u_1 - u_2)^2 = (p_2 - p_1)\left(\frac{1}{\rho_1} - \frac{1}{\rho_2}\right),$$

where, as usual, suffix 1 is ahead of the shock and suffix 2 is behind.
From (6.24) and (6.25) show that

$$\frac{\rho_2}{\rho_1} = \frac{(\gamma - 1)p_1 + (\gamma + 1)p_2}{(\gamma + 1)p_1 + (\gamma - 1)p_2}.$$

Now suppose that a normal shock with pressure ratio $\pi(= p_2/p_1)$ is reflected from a plane parallel rigid boundary. Show that the pressure ratio of the reflected shock is p, where

$$p = \frac{(3\gamma - 1)\pi - (\gamma - 1)}{(\gamma - 1)\pi + (\gamma + 1)}.$$

6.5 A plane shock wave is reflected from a parallel plane wall. Show that if U_1 is the speed of approach and U_2 the speed of departure, then

$$(U_1 - u)^2 - c^2 = -\left(\frac{\gamma + 1}{2}\right)u(U_1 - u),$$

$$(U_2 + u)^2 - c^2 = \frac{\gamma + 1}{2}u(U_2 + u),$$

where u, c are the velocity and speed of sound behind the incoming shock. Deduce that

$$U_1 > U_2$$

as long as $\gamma < 3$.

*6.6 You are given the following model for one-dimensional viscous compressible gas dynamics, where the velocity has been made nondimensional with c_0, the speed of sound in the undisturbed gas:

$$\frac{\partial \rho}{\partial t} + \frac{\partial}{\partial x}(\rho u) = 0,$$

$$\gamma \rho \left(\frac{\partial u}{\partial t} + u \frac{\partial u}{\partial x} \right) + \frac{\partial p}{\partial x} = \delta \frac{\partial^2 u}{\partial x^2},$$

$$\frac{\partial p}{\partial t} + u \frac{\partial p}{\partial x} - \frac{\gamma p}{\rho} \left(\frac{\partial \rho}{\partial t} + u \frac{\partial \rho}{\partial x} \right) = \delta(\gamma - 1) \left(\frac{\partial u}{\partial x} \right)^2,$$

where δ is a measure of the viscosity of the gas and is small.

An impulsive piston travels at constant speed $\varepsilon \ll 1$ into a tube containing the gas at rest with $p = 1$ and $\rho = 1$. Use (6.30) to show that, when $\delta = 0$ a shock travels ahead of the piston with dimensionless speed $1 + O(\varepsilon)$.

To understand the structure of a viscous shock, change to variables $\varepsilon \xi = x - t$ and t and write $u = \varepsilon \bar{u}$, $\rho = 1 + \varepsilon \bar{\rho}$, $p = 1 + \varepsilon \bar{p}$ in the above equations. Assume $\delta = O(\varepsilon^2)$ and show that, to lowest order in ε,

$$\bar{p}_\xi = \gamma \bar{u}_\xi = \gamma \bar{\rho}_\xi,$$

but that $\bar{p}, \bar{u}, \bar{\rho}$ are not defined uniquely at this order. Now keep terms of $O(\varepsilon)$ and show that

$$-\bar{\rho}_\xi + \bar{u}_\xi = -\varepsilon(\bar{\rho}_t + (\bar{\rho}\bar{u})_\xi),$$
$$-\gamma \bar{u}_\xi + \bar{p}_\xi = \varepsilon(-\gamma \bar{u}_t - \gamma \bar{u}\bar{u}_\xi + \gamma \bar{\rho}\bar{u}_\xi + \tfrac{\delta}{\varepsilon^2} \bar{u}_{\xi\xi}),$$
$$-\bar{p}_\xi + \gamma \bar{u}_\xi = -\varepsilon(\bar{p}_t + \bar{u}\bar{p}_\xi + \gamma \bar{p}\bar{u}_\xi).$$

Hence, deduce that $\bar{u}$ satisfies Burgers equation

$$\frac{\partial \bar{u}}{\partial t} + \frac{\gamma + 1}{2} \bar{u} \frac{\partial \bar{u}}{\partial \xi} = \frac{\delta}{2\varepsilon^2} \frac{\partial^2 \bar{u}}{\partial \xi^2}.$$

Explain why appropriate boundary conditions are $\bar{u} \to 1$, as $\xi \to -\infty$ and $\bar{u} \to 0$ as $\xi \to +\infty$. Show that there is a travelling wave solution such that $\bar{u} = U(\xi - \lambda t)$ which satisfies these boundary conditions as long as $\lambda = (\gamma + 1)/4$. Why would this not be possible if the boundary condition at $\xi \to -\infty$ was $\bar{u} = -1$?

R 6.7 Rearrange (6.46) to show that

$$\tan \theta = 2 \cot \beta \frac{M_1^2 \sin^2 \beta - 1}{(\gamma + \cos 2\beta)M_1^2 + 2}.$$

Show that $\theta = 0$ when $\beta = \pi/2$ and when $\beta = \sin^{-1}(1/M_1)$. Deduce that, for fixed M_1, $\theta(\beta)$ has at least one maximum (which can be shown to be close to, but not quite equal to, the value of β for which $M_2 = 1$). Show also that, when θ is small and β is near $\sin^{-1}(1/M_1)$,

$$\theta \sim \frac{4(M_1^2 - 1)}{(\gamma + 1)M_1^2}(\beta - \sin^{-1}\frac{1}{M_1}).$$

Deduce that, for such weak shocks, the downstream pressure, Mach number and velocity are given by

$$\frac{p_2}{p_1} \sim 1 + \gamma M_1^2 \theta / \sqrt{M_1^2 - 1},$$

$$\frac{M_2}{M_1} \sim 1 - \theta(1 + \frac{\gamma - 1}{2}M_1^2) / \sqrt{M_1^2 - 1}$$

and

$$\frac{u_2}{u_1} \sim 1 - \frac{\theta}{\sqrt{M_1^2 - 1}},$$

respectively, where u_1 is the velocity in front of the shock and u_2 is the component of the velocity in the same direction behind the shock.

*6.8 A gas with speed u_1 and sound speed c_1 flows homentropically to a state, where the Mach number is unity and its sound speed is c_*, which is called the *critical speed of sound*. Show that

$$(\gamma + 1)c_*^2 = (\gamma - 1)u_1^2 + 2c_1^2$$

and hence use (6.42) to show that the critical speeds of sound on either side of an oblique shock are the same.

Suppose a gas with velocity $(u_1, 0)$ and sound speed c_1 passes through a shock, inclined at an angle β to the x-axis, to a state with velocity (u_2, v_2). Show that

$$u_1 \cos \beta = u_2 \cos \beta + v_2 \sin \beta,$$

$$\rho_1 u_1 \sin \beta = \rho_2(u_2 \sin \beta - v_2 \cos \beta),$$

$$p_1 + \rho_1 u_1^2 \sin^2 \beta = p_2 + \rho_2(u_2 \sin \beta - v_2 \cos \beta)^2$$

and

$$\frac{1}{2}u_1^2 \sin^2 \beta + \frac{c_1^2}{\gamma - 1} = \frac{1}{2}(u_2 \sin \beta - v_2 \cos \beta)^2 + \frac{c_2^2}{\gamma - 1}$$

$$= \frac{\gamma + 1}{2(\gamma - 1)}c_*^2 - \frac{1}{2}u_1^2 \cos^2 \beta.$$

Eliminate p_1, ρ_1, p_2, ρ_2 and β from these equations to show that

$$v_2^2 = \frac{(u_1 - u_2)^2 (u_1 u_2 - c_*^2)}{([2/(\gamma + 1)]u_1^2 + c_*^2 - u_1 u_2)}.$$

Sketch this curve in the (u_2, v_2) plane for given u_1, c_*, and show that only the segment $c_*^2/u_1 < u_2 < u_1$ is physically relevant. Confirm that there are two values of (u_2, v_2) for any given value of v_2/u_2 less than the maximum possible deflection.

The curve in the (u_2, v_2) plane is called the *shock polar*.

6.9 Show that if the Prandtl–Meyer expansion (5.74)–(5.75) is weak, the flow deflection θ, which is now negative, is approximately related to the downstream Mach number M_2 by

$$\theta \sim (\mu_1 - \mu)f'(\mu_1)$$

$$\sim \frac{(M_2 - M_1)}{M_1\sqrt{M_1^2 - 1}} \left(1 - \frac{(\gamma + 1)M_1^2}{2 + (\gamma - 1)M_1^2}\right).$$

Deduce that

$$\frac{M_2}{M_1} \sim 1 - \theta(1 + \frac{\gamma - 1}{2}M_1^2)/\sqrt{M_1^2 - 1},$$

which is the same as the formula obtained for a weak shock in Exercise 6.7. However, remember θ is now negative and M_2 is now greater than M_1 and that weak shocks and weak expansions refract the flow in *opposite* directions.

6.10 A circular cone is placed in a uniform supersonic stream with its axis parallel to the stream. Show that if the resulting shock wave is a concentric circular cone, then there is a velocity potential Φ such that the radial and transverse velocities between the shock and the body are

$$u_r = \frac{\partial \Phi}{\partial r}, \quad u_\theta = \frac{1}{r}\frac{\partial \Phi}{\partial \theta},$$

where r, θ are spherical polar coordinates. Given that

$$\nabla.(\rho\mathbf{u}) = \frac{1}{r\sin\theta}\frac{\partial}{\partial\theta}(\sin\theta\rho u_\theta) + \frac{1}{r^2}\frac{\partial}{\partial r}(r^2\rho u_r),$$

show that there is a similarity solution in which $\Phi = r\phi(\theta)$ with

$$\left(1 - \frac{1}{c^2}\left(\frac{d\phi}{d\theta}\right)^2\right)\frac{d^2\phi}{d\theta^2} - \left(\frac{\phi}{c^2}\frac{d\phi}{d\theta} - \cot\theta\right)\frac{d\phi}{d\theta} + 2\phi = 0,$$

where
$$c^2 = -\frac{(\gamma - 1)}{2}\left(\phi^2 + \left(\frac{d\phi}{d\theta}\right)^2\right) + \text{constant}.$$

R6.11 Write $F_i = (u_i - \dot{X}_s)/\sqrt{g\eta_i}$ in (6.49)–(6.50) and show that

$$[s^3 F] = [\frac{1}{2}s^4 + F^2 s^4] = 0.$$

Deduce that $F_1^2 = s_2^2(1 + (s_2^2/s_1^2))/2s_1^2$ and use (6.52) to infer that $F_1 \geq 1$ and $F_2 \leq 1$.

Note the analogy between these inequalities for the Froude number in shallow water theory and the inequalities (6.28) for the Mach number in gas dynamics.

6.12 A bore invades water originally at rest in a straight horizontal channel of uniform rectangular cross-section. The depth of the water increases from H to $2H$ by the passage of the bore. Show that the velocity behind the bore is $\sqrt{3gH}$. The bore is reflected at the closed end of the channel. Show that after reflection the depth of water at the closed end is $\frac{1}{2}(1 + \sqrt{33})H$.

*6.13 Show that for the two-dimensional shallow water equations of Exercise 5.5, the steady shock (bore) relations for the conservation of mass and momentum are

$$\frac{dy}{dx} = \frac{[\eta v]}{[\eta u]} = \frac{[\eta u v]}{[\frac{1}{2}g\eta^2 + \eta u^2]} = \frac{[\frac{1}{2}g\eta^2 + \eta v^2]}{[\eta u v]}.$$

Show that the energy dissipation across the bore depends on dy/dx and use the result of Exercise 5.5 to deduce that if a uniform stream encounters a curved bore, the downstream vorticity $\partial v/\partial x - \partial u/\partial y$ will be non-zero.

R 6.14 Water of depth s_l^2/g is contained in $-\infty < x < 0$ and is separated by a sluice gate from water of depth s_r^2/g in $0 < x < \infty$, where $s_r < s_l$. At time $t = 0$, the sluice gate is suddenly removed. Show that the solution comprises the following:

(i) An expansion fan in $-s_l t < x < (u_1 - s_1)t$.
(ii) A region of uniform flow, where $s = s_1, u = u_1$ for $(u_1 - s_1)t < x < Vt$.
(iii) A hydraulic jump at $x = Vt$.

Write down sufficient equations to determine u_1, s_1 and V, and show that if $u_1 > 2s_l/3$, and $t > 0$, then the water depth at $x = 0$ is $4s_l^2/9g$, and the discharge rate is $8s_l^3/27g$.

R 6.15 Gas flows steadily out of a reservoir, where the density is ρ_0 and the sound speed c_0, into a duct of slowly varying cross-section $A(x)$. The duct area initially decreases to a minimum at $x = X$ and then increases. Show that if the Mach number is M and the mass flow in the duct is Q, then

$$\frac{\rho_0 c_0}{Q} \frac{dA}{dx} = (1 - \frac{1}{M^2})\left(1 + (\frac{\gamma - 1}{2})M^2\right)^{(3-\gamma)/2(\gamma-1)} \frac{dM}{dx}.$$

Deduce that if the duct is choked so that $M = 1$ at $x = X$, then

$$Q = \left(\frac{2}{\gamma + 1}\right)^{(\gamma+1)/2(\gamma-1)} \rho_0 c_0 A(X).$$

6.16 Water (with unit density) flows into an open horizontal channel from a reservoir where the total head is gH. Show that if the channel breadth $b(x)$ decreases to a minimum b_* downstream before increasing again, then if the Froude number F attains the value unity, it does so when $b = b_*$. Prove that in such a choked flow, the flow rate is

$$q = (\frac{2}{3})^{3/2} g^{1/2} H^{3/2} b_*.$$

6.17 (i) Show that the density and pressure ratios across the expansion fan generated by a piston moved impulsively out of a tube with velocity U_p, as in Exercise 5.17, are given by

$$\frac{\rho_2}{\rho_1} = \left(1 - \frac{\gamma - 1}{2}\frac{|U_p|}{c_1}\right)^{2/(\gamma-1)}, \quad \frac{p_2}{p_1} = \left(\frac{\rho_2}{\rho_1}\right)^\gamma,$$

where $|U_p|$ is the piston speed, assumed less than $2c_1/(\gamma - 1)$, and subscripts 1 and 2 refer to conditions ahead of and behind the fan, respectively.

(ii) Inviscid gas is contained in an infinite shock tube lying along the x axis. An impermeable membrane at $x = 0$ separates gas with pressure p_l and sound speed c_l in $x < 0$ from the same gas at conditions p_r, c_r in $x > 0$, where $p_l > p_r$. At time $t = 0$ the membrane is ruptured. Show that the subsequent flow comprises

(a) An expansion fan in $-c_l t < x < (V - c_2)t$, where V is the speed of the contact discontinuity.

(b) A uniform flow region in $(V - c_2)t < x < Vt$, in which $u = V$, $p = p_1$, $c = c_2$.

(c) A uniform flow region in $Vt < x < Ut$, in which $u = V$, $p = p_1$, $c = c_1$.

(d) A shock at $x = Ut$, where the unknowns satisfy

$$c_2 = c_l - \frac{\gamma - 1}{2}V, \quad \frac{p_1}{p_l} = \left(1 - \frac{(\gamma - 1)}{2}\frac{V}{c_l}\right)^{2\gamma/(\gamma-1)},$$

$$(U - V)\rho_1 = U\rho_r,$$

$$p_1 + \rho_1(U - V)^2 = p_r + \rho_r U^2$$

and

$$\frac{c_1^2}{\gamma - 1} + \frac{1}{2}(V - U)^2 = \frac{c_r^2}{\gamma - 1} + \frac{1}{2}U^2,$$

where ρ_r and ρ_1 are the densities ahead of and behind the shock. Show that

$$\frac{V}{c_l} = \frac{2}{\gamma - 1}\left(1 - \left(\frac{p_1}{p_l}\right)^{(\gamma-1)/2\gamma}\right),$$

and that

$$\frac{V}{c_r} = \left(\frac{p_1}{p_r} - 1\right)\left(\frac{2/\gamma}{(\gamma + 1)p_1/p_r + (\gamma - 1)}\right)^{1/2},$$

and hence deduce the *shock tube equation*

$$\frac{p_l}{p_r} = \frac{p_1}{p_r}\left(1 - \frac{(\gamma - 1)(c_r/c_l)(p_1/p_r - 1)}{[2\gamma(\gamma + 1)p_1/p_r + \gamma - 1]^{1/2}}\right)^{-2\gamma/(\gamma-1)}.$$

Which of the flow variables are continuous at the contact discontinuity $x = Vt$?

6.18 Suppose two unequal weak shocks make angles β, β' with a stream of Mach number M_1 as in Figure 6.12. Show that the deflections satisfy

$$\theta - \phi \sim -\theta' + \phi',$$

and use the results of Exercise 6.7 to show that

$$\theta + \phi \sim \theta' + \phi'.$$

Show that the flow deflection satisfies

$$\theta - \theta' \sim \frac{4(M_1^2 - 1)}{(\gamma + 1)M_1^2}(\beta - \beta'),$$

and that, in general, there will be a contact discontinuity (in this case a vortex sheet) separating the downstream region into parallel gas streams with unequal speeds.

6.19 Suppose that, instead of the configurations in Figure 6.12, two weak shocks intersect as in Figure 6.13. Show that the shocks can merge to form a third shock with $\phi \sim \theta + \theta'$, and that there will again be a contact discontinuity in the downstream flow.

6.20 Use (6.46) to show that

$$\tan \theta = \frac{2(M_1^2 \sin^2 \beta_1 - 1) \cot \beta_1}{2 + \gamma M_1^2 + M - 1^2 \cos 2\beta_1}.$$

Hence, show that if $M_1^2 \sin^2 \beta_1 = 1 + \varepsilon$ where $\varepsilon \ll 1$, then

$$\theta \sim \frac{2\varepsilon \cos \beta_1 \sin \beta_1}{\gamma + 1}.$$

Using the notation of Figure 6.14 to consider the reflection of a weak shock, show from (6.45) that

$$M_2^2 \sin^2 \beta_1 \sim 1 + \varepsilon \left(\frac{4 \cos^2 \beta_1}{\gamma + 1} - 1 \right).$$

By considering the flow deflection at the reflected shock, show that

$$M_2^2 \sin^2 \beta_2 \sim 1 + \varepsilon$$

and hence show that

$$\beta_3 \sim \beta_1 + \varepsilon \left(\tan \beta_1 - \frac{4 \cos \beta_1 \sin \beta_1}{(\gamma + 1)} \right).$$

6.21 A weak shock I impinges on a vortex sheet which separates two supersonic streams with Mach numbers M_1 and M_1' as shown in Figure 6.16.
Show that if the deflections θ, ϕ, θ' are measured as shown in Figure 6.16, then, using the results of Exercise 6.7,

$$\theta - \phi \sim \theta'$$

and

$$(\theta + \phi) \frac{M_1^2}{\sqrt{M_1^2 - 1}} \sim \theta' \frac{M_1'^2}{\sqrt{M_1'^2 - 1}}.$$

Show further that there will be a contact discontinuity in the downstream flow.

If $M_1 > M'_1 > \sqrt{2}$, show that $\phi < 0$, so that the second shock R above the vortex sheet will be replaced by an expansion fan for which the above results still apply (see Exercise 6.9). Show that the strength of the shock T transmitted by the vortex sheet is always less than the strength of the incident shock I.

*6.22 Suppose gas flows steadily, with inlet velocity U, down a slowly varying channel of length L with walls given by

$$y = \pm S(\varepsilon x),$$

where $\varepsilon \ll 1$. Assuming that $u = O(U)$, $v = O(\varepsilon U)$, $x = O(L)$ and variations in y are of $O(\varepsilon L)$, show that the equations of continuity and momentum are approximated by

$$\frac{\partial}{\partial x}(\rho u) + \frac{\partial}{\partial y}(\rho v) = 0,$$

$$\frac{\partial}{\partial x}(\rho u^2) + \frac{\partial}{\partial y}(\rho u v) + \frac{\partial p}{\partial x} = 0,$$

$$\frac{\partial p}{\partial y} = 0,$$

with

$$v = \pm \varepsilon u S' \quad \text{on } y = \pm S(\varepsilon x).$$

Show that

$$\frac{d}{dx} \int_{-S}^{S} \rho u \, dy = 0$$

and

$$\frac{d}{dx} \int_{-S}^{S} \rho u^2 \, dy + 2S \frac{dp}{dx} = 0.$$

Assuming additionally that the flow is, to lowest order, irrotational and homentropic, show that u, ρ and p are all approximately functions of x alone and that their averages over the channel width satisfy

$$\bar\rho \bar u S = \text{constant}, \quad \frac{1}{2}\bar u^2 + \frac{\gamma \bar p}{(\gamma - 1)\bar\rho} = \text{constant}, \quad \bar p / \bar\rho^\gamma = \text{constant},$$

where $\bar\rho = (1/2S) \int_{-S}^{S} \rho \, dy$ and similarly for $\bar u$ and $\bar p$.

*6.23 (i) Show that the far field of a supersonic stream past a thin wing is modelled, in $Y > 0$, by

$$\frac{\partial u}{\partial Y} + \frac{\gamma + 1}{2B} M^4 u \frac{\partial u}{\partial \xi} = 0$$

in the notation of Section 6.3.2, with $u = -(1/B)f'(\xi)$ on $Y = 0$. Show also that the Rankine–Hugoniot condition for this equation is that the leading edge shock slope in (ξ, Y) coordinates is

$$\frac{d\xi}{dY} = \frac{(\gamma + 1)M^4}{4B} \frac{[u^2]}{[u]} = \frac{(\gamma + 1)M^4}{4B} u_+,$$

where u_+ is the value of u just downstream of the leading shock. Check this result by using Exercise 6.7 to show that

$$\varepsilon u_+ = -\frac{4B}{(\gamma + 1)M^2} \left(\beta - \sin^{-1} \frac{1}{M} \right),$$

where $d\xi/dY = (1/\varepsilon)(\cot \beta - B)$.

(ii) Suppose the wing is such that $f(\xi) = l^2 - \xi^2$ for $-l < \xi < l$. Show that the leading edge shock wave is given by

$$\frac{d\xi}{dY} = \frac{(\gamma + 1)M^4}{4B} u_0(\xi_0),$$

where $u_0(\xi_0)$ is the value of u on the characteristic

$$\frac{d\xi}{dY} = \frac{(\gamma + 1)M^4}{2B} u_0(\xi_0),$$

with $\xi = \xi_0$ when $Y = 0$. Deduce that

$$\xi_0 = \frac{\xi}{(1 + ((\gamma + 1)M^4/B)Y)}$$

and hence show that the shock wave is the parabola

$$\frac{(\gamma + 1)M^4 Y}{B} + 1 = \alpha \xi^2$$

for some constant α. Consider the flow as $Y \downarrow 0$ to show that $\alpha = 1/l$. A similar calculation reveals the existence of a parabolic shock from the trailing edge, so that the far field pressure is an "N-wave".

*6.24 (i) Show that if β and θ are small, (6.44) implies that

$$\theta \sim \frac{2\beta}{\gamma + 1}\left(1 - \frac{1}{M_1^2 \beta^2}\right).$$

(ii) In hypersonic flow, the variables $(\bar{p}, \bar{\rho}, \bar{v}, \bar{x}, \bar{y})$ in (6.66), (6.68) and (6.69) are identified with the variables (p, ρ, u, t, x) in the one-dimensional unsteady gas dynamic equations (5.3)–(5.5).

If a piston is pushed into a gas at rest with pressure p_1 and density ρ_1, show from (6.24)–(6.25) that the pressure p_2, velocity u_2 and density ρ_2 just behind the shock, whose position is given by $x = \dot{X}_s(t)$, are given by

$$\frac{p_2}{p_1} = \frac{2\gamma M_1^2 - (\gamma - 1)}{(\gamma + 1)}, \quad \frac{\rho_2}{\rho_1} = \frac{(\gamma + 1)M_1^2}{2 + (\gamma - 1)M_1^2},$$

$$u_2 = \frac{2\dot{X}_s}{\gamma + 1}\left(1 - \frac{1}{M_1^2}\right),$$

where $M_1 = \dot{X}_s/c_1$ and $c_1^2 = \gamma p_1/\rho_1$.

For the hypersonic problem, show that if the shock is given by $\bar{y} = Y_s(\bar{x})$, then $\beta = \varepsilon Y_s'$, and, from (6.43), (6.44), the shock relations are

$$\frac{p_2}{p_1} = \frac{2\gamma M_1^2 \beta^2 - (\gamma - 1)}{\gamma + 1} \quad \text{and} \quad \frac{\rho_2}{\rho_1} = \frac{(\gamma + 1)M_1^2 \beta^2}{2 + (\gamma - 1)M_1^2 \beta^2};$$

deduce from (i) above that

$$\bar{v} = \frac{2Y_s'}{\gamma + 1}\left(1 - \frac{1}{M_1^2 \beta^2}\right).$$

Hence, complete the identification that leads to the principle of hypersonic similitude.

*6.25 Show that if gas streams with Mach number M past a thin wing, with slope $O(\varepsilon)$ and $M\varepsilon \gg 1$, then, in the notation of (6.66)–(6.69) and using the results of Exercise 6.24, the shock conditions of $\bar{y} = \bar{Y}_s(\bar{x})$ are

$$\bar{p} = \frac{2}{\gamma + 1}\bar{Y}_s'^{1/2}, \quad \bar{\rho} = \frac{\gamma + 1}{\gamma - 1} \quad \text{and} \quad \bar{v} = \frac{2\bar{Y}_s'}{\gamma + 1}.$$

Deduce that if the wing is $\bar{y} = b\bar{x}^k$ for $\bar{x} > 0$, then there is a similarity solution

$$\bar{Y}_s = s\bar{x}^k, \quad \bar{p} = \bar{x}^{2(k-1)}P(\zeta), \quad \bar{\rho} = R(\zeta), \quad \bar{v} = \bar{x}^{k-1}V(\zeta),$$

where $\zeta = \bar{y}/\bar{x}^k$, which satisfies

$$-k\gamma R' + (RV)' = 0,$$

$$(k-1)V - k\gamma V' + VV' = -P'/R$$

and

$$\left(2k - 2 + (V - k\zeta)\frac{d}{d\zeta}\right)(P/R^\gamma) = 0,$$

with

$$P(s) = \frac{2k^2 s^2}{\gamma + 1}, \quad R(s) = \frac{\gamma + 1}{\gamma - 1}, \quad V(s) = \frac{2ks}{\gamma + 1}$$

and

$$V(b) = kb.$$

*6.26 The equation of a ship moving in the z-direction with velocity V is given by
$F(x, y, z - Vt) = 0$. Show that in steady flow, with $\xi = z - Vt$, the potential
for the flow generated by the passage of the ship satisfies

$$\frac{\partial^2 \phi}{\partial x^2} + \frac{\partial^2 \phi}{\partial y^2} + \frac{\partial^2 \phi}{\partial \xi^2} = 0,$$

with

$$-V\frac{\partial \phi}{\partial \xi} + g\eta + \frac{1}{2}\left(\left(\frac{\partial \phi}{\partial x}\right)^2 + \left(\frac{\partial \phi}{\partial y}\right)^2 + \left(\frac{\partial \phi}{\partial \xi}\right)^2\right) = 0$$

and

$$\frac{\partial \phi}{\partial y} = -V\frac{\partial \eta}{\partial \xi} + \frac{\partial \phi}{\partial x}\frac{\partial \eta}{\partial x} + \frac{\partial \phi}{\partial \xi}\frac{\partial \eta}{\partial \xi}$$

on the free surface $y = \eta$, and

$$V\frac{\partial F}{\partial \xi} = \frac{\partial \phi}{\partial x}\frac{\partial F}{\partial x} + \frac{\partial \phi}{\partial y}\frac{\partial F}{\partial y} + \frac{\partial \phi}{\partial \xi}\frac{\partial F}{\partial \xi}$$

on the ship $F(x, y, \xi) = 0$; also $|\nabla\phi| \to 0$ at infinity since there are no
incoming waves.

Now suppose the ship is narrow and of length l, so that
$F(x, y, \xi) = \tilde{F}(X, Y, \xi)$ where $x = \varepsilon lX$, $y = \varepsilon lY$, $\xi = l\zeta$. Also suppose
that the Froude number is so large that $\varepsilon V^2/gl = f = O(1)$ as $\varepsilon \to 0$. Show
that if $\phi = \varepsilon^2 lV\tilde{\phi}$, $\eta = \varepsilon l\tilde{\eta}$, then, to lowest order in ε,

$$\frac{\partial^2 \tilde{\phi}}{\partial X^2} + \frac{\partial^2 \tilde{\phi}}{\partial Y^2} = 0,$$

with, on $Y = \tilde{\eta}$,

$$-\frac{\partial\tilde{\phi}}{\partial\zeta} + f^{-1}\tilde{\eta} + \frac{1}{2}((\frac{\partial\tilde{\phi}}{\partial X})^2 + (\frac{\partial\tilde{\phi}}{\partial Y})^2) = 0$$

and

$$\frac{\partial\tilde{\phi}}{\partial Y} = -\frac{\partial\tilde{\eta}}{\partial\zeta} + \frac{\partial\tilde{\phi}}{\partial X}\frac{\partial\tilde{\eta}}{\partial X},$$

and, on $\tilde{F} = 0$,

$$\frac{\partial\tilde{F}}{\partial\zeta} = \frac{\partial\tilde{\phi}}{\partial X}\frac{\partial\tilde{F}}{\partial X} + \frac{\partial\tilde{\phi}}{\partial Y}\frac{\partial\tilde{F}}{\partial Y}.$$

Show that, when $-\zeta$ is identified with time t, these are the equations of surface gravity waves in two dimensions driven by a surface penetrating wavemaker $\tilde{F}(X, Y, -t) = 0$.

6.27 Use the chain rule to show that

$$\rho\left(\frac{dX}{dt} - u\right) = \rho_0 \frac{da}{dt}, \qquad (*)$$

where X, t are Eulerian coordinates and a, t are Lagrangian coordinates. Use (6.16) to show that the left-hand side of $(*)$ is continuous even at a shock wave.

6.28 Show that when $\hat{n}$ satisfies (6.74), its Fourier transform

$$\bar{n}(k, t) = \int_{-\infty}^{\infty} \hat{n}e^{ikx}\, dk$$

satisfies

$$\frac{d^2\bar{n}}{dt^2} + \frac{k^2}{1 + k^2}\bar{n} = 0.$$

Assuming the plasma is initially at rest with $\hat{n}$ given by (6.75), and recalling (3.59), show that

$$\bar{n} = \frac{1}{ik}\cos\left(\frac{kt}{\sqrt{1 + k^2}}\right)$$

as long as Im $k > 0$.

The Fourier inversion theorem then implies that

$$\hat{n} = \frac{1}{2\pi i}\int_{-\infty + i\varepsilon}^{\infty + i\varepsilon} \bar{n}e^{-ikx}\, dk,$$

where $\varepsilon > 0$. By splitting the integral into Rl $k \gtrless 0$, and accounting for the region near $k = 0$ as $\varepsilon \to 0$, deduce (6.76).

Chapter 7
Epilogue

As explained in Chapter 1, the authors have changed the emphasis from that of its progenitor "Inviscid Fluid Flows" by reorganising the material so that the applicability of the analysis can be demonstrated as widely as possible. This has revealed how many mathematical methods that seem to be intimately connected with compressible flow or shallow water theory turn out to be equally useful in areas such as electromagnetism, plasma physics and plasticity.

However, this is by no means the end of the story as far as the mathematical theory of wave propagation is concerned. In recent decades, there has been a spectacular blossoming of theory associated with travelling disturbances in chemical and biological systems as distinct from mechanical or electromagnetic ones that we discuss here. These "less classical" waves can exhibit many of the features, such as steepening, dispersion, reflection and diffraction, that we have encountered in the preceding pages. However, there is something distinctive about waves that are governed by equations whose linearisation yields a dispersion relation such that ω is real for all real k. This is true for the *high frequency* behaviour of every *hyperbolic* system. This means that these systems have a robustness like that of simple harmonic motion in that they can exist without relying on any input or loss from their surroundings. As we have seen in Chapter 4, the theory of linear wave propagation in a non-dissipative medium is very powerful and can be very complicated mathematically. This is not surprising when one considers that phenomena as diverse as scattering, diffraction and focusing can all be predicted in both the frequency and the time domains.

One of the main themes of this book has been to show how the presence of non-linear terms that are small almost everywhere can have a profound effect on the robustness of linear non-dissipative waves. If the linear model is non-dispersive, as is the case for acoustic or plastic wave propagation, even small nonlinearity can lead to discontinuous solutions or shock waves. However, it is also possible that a balance between dispersion and weak nonlinearity can give rise to waves that are able to propagate indefinitely as in the theory for solitons or resonance. Moreover, we have seen from the plasma models that when charged particles have even a small

© Springer Science+Business Media New York 2015
H. Ockendon, J.R. Ockendon, *Waves and Compressible Flow*,
Texts in Applied Mathematics 47, DOI 10.1007/978-1-4939-3381-5_7

temperature, then random effects can cause waves to decay in time. Unsurprisingly, strong temporal decay is present for waves in electric conductors and dissipation is a mechanism that can smooth the shock waves that arise in inviscid gases. Indeed, dissipation provides a reassuring physical basis for the mathematical theory of weak solutions though we also note that dissipative effects can engender temporal instability as evinced by Tollmein–Schlichting waves.

It is a testimony to the power of mathematics that it is able not only to illuminate a fascinating variety of wave behaviour exhibited by fluids but can also shed light on waves in so many other areas of applied science.

Appendix

A.1 Heat Engines and Entropy

A knowledge of gas dynamic piston problems is useful in understanding the theory of *heat engines*, which are devices for converting thermal energy into mechanical energy. The most famous result concerning the efficiency of such engines considers the processing of a gas in an extensible cylinder aligned with the x-axis. The cylinder is initially immersed in a hot bath at temperature T_H and then the following four stages take place.

(i) The cylinder is slowly extended, while still in the hot bath, to allow the gas to expand isothermally at temperature T_H to a lower pressure p_H. From the perfect gas law (2.13), this means that p/ρ is constant during this expansion. Thus, the gas behaves as if its value of γ was unity as it changes from state A to state B in Figure A.1. During this process, the gas will satisfy the equations

$$\left(\frac{\partial}{\partial t} + (u \pm c_0)\frac{\partial}{\partial x}\right)(u \pm c_0 \log \rho) = 0,$$

where $p/\rho = c_0$, rather than (5.9). Moreover, since $de/dt = c_v dT/dt = 0$, the energy equation (2.17) reduces to

$$\rho\frac{dQ}{dt} + \frac{p}{\rho}\frac{d\rho}{dt} = 0, \tag{A.1}$$

where the local heat flux from the bath is $Q(t)$.[1] Thus, the heat flux from the bath to the cylinder is precisely equal to the work done by the pressure; no external mechanical work is needed to maintain isothermal conditions. From (2.20) with $T = T_H$, we see that there is a net entropy change $S_H = Q_H/T_H$, where Q_H is the net heat flux and this gives us an insight into a physical interpretation of entropy.

---

[1] A model that incorporates the inertia of the gas is given in Exercise 3.6.

© Springer Science+Business Media New York 2015
H. Ockendon, J.R. Ockendon, *Waves and Compressible Flow*,
Texts in Applied Mathematics 47, DOI 10.1007/978-1-4939-3381-5

(ii) The cylinder is now removed from the bath and thermally insulated before the gas inside is allowed to slowly expand again to a lower pressure p_C and temperature T_C. This part of the process is controlled by a piston at the end of the cylinder which will do mechanical work W as the state of the gas changes from B to C in Figure A.1. This adiabatic expansion is a piston withdrawal problem as described in Section 5.4.1 after taking the presence of the end wall into account.

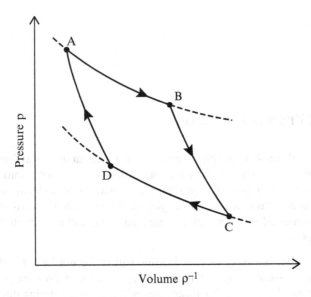

Fig. A.1 The Carnot cycle

(iii) The thermal insulation is removed from the cylinder before it is slowly compressed isothermally in a cold bath at temperature T_C, thereby increasing the pressure and density and rejecting heat Q_C to the bath and going from point C to D in Figure A.1. As in (i) no mechanical work will be done.

(iv) The cycle is closed by removing the cylinder from the cold bath, reinsulating it and slowly compressing the gas by doing work on the piston until the temperature T_H is reached. The key requirement for this process to be a *Carnot Cycle*[2] is that this work must be the same in magnitude as W, the work produced in (ii), and the gas will then have returned to state A. We recall the cautionary remark in Section 5.4.1 about the possibility of shock waves occurring during the compressive phase and this is why the compression must be done slowly enough for there to be no shock, and hence no corresponding entropy changes.

The efficiency η of a heat engine is defined as the ratio of W, the mechanical work created, to Q_H, the net heat input. However, from the first law of thermodynamics, $W = Q_H - Q_C$ and so

---

[2] If the cycle in Figure A.1 is reversed, it is called a *Carnot refrigerator*.

$$\eta = 1 - \frac{Q_C}{Q_H}. \tag{A.2}$$

Using the fact that heat cannot flow from a cold region to a hot one, it can be shown that, if the mechanical work in (iv) is not W, then η is decreased so the Carnot cycle is maximally efficient. Since there is no dissipation in the Carnot cycle, the entropy changes in (i) and (iii) are equal and opposite[3] so that, from (2.20),

$$\frac{Q_C}{T_C} = \frac{Q_H}{T_H} \quad \text{and} \quad \eta = 1 - \frac{T_C}{T_H}.$$

Thus, it is plausible that every heat engine has efficiency less than unity and that the efficiency only tends to unity as T_C tends to absolute zero. It is thus possible to interpret absolute zero as the temperature at which a heat engine rejects no heat into the cold bath in stage (iii), see [55] for more details.

---

[3] Note that the Second Law of Thermodynamics only says that entropy cannot decrease *globally*.

References

1. Acheson, D. J. (1990). *Elementary fluid dynamics*. Oxford: Oxford University Press.
2. Anderson, J. D. (1989). *Hypersonic and high temperature gas dynamics*. New York: McGraw-Hill.
3. Arscott, F. M. (1964). *Periodic differential equations. An introduction to Mathieu, Lamé and allied functions*. Elmsford, NY: Pergamon.
4. Benjamin, T. B., & Ursell, F. (1954). The stability of the plane free surface of a liquid in vertical periodic motion. *Proceedings of the Royal Society A, 225*, 505–515.
5. Billingham, J., & King, A. C. (2000). *Wave motion*. Cambridge: Cambridge University Press.
6. Born, M., & Wolf, E. (1980). *Principles of optics* (6th edn.). Elmsford, NY: Pergamon.
7. Chapman, C. J. (2000). *High speed flow*. Cambridge: Cambridge University Press.
8. Chapman, S., & Cowling, T. G. (1952). *The mathematical theory of nonuniform gases*. Cambridge: Cambridge University Press.
9. Chapman, S. J., Lawry, J. M. H., Ockendon, J. R., & Tew, R. H. (1999). On the theory of complex rays. *SIAM Review, 41*, 417–509.
10. Chen, F. F. (1984). *Introduction to plasma physics and controlled fusion*. Volume 1: plasma physics. New York: Springer.
11. Chester, C. R. (1971). *Techniques in partial differential equations*. New York: McGraw-Hill.
12. Coulson, C. A., & Boyd, T. J. M. (1979). *Electricity*. London: Longman.
13. Courant, R., & Friedrichs, K. O. (1948). *Supersonic flow and shock waves*. New York: Interscience.
14. Courant, R., & Hilbert, D. (1962). *Methods of mathematical physics* (Vol. I). New York: Interscience.
15. Davison, L. (2010). *Fundamentals of shock wave propagation in solids*. New York: Springer.
16. Dodd, R. K., Eilbeck, J. C., Gibbon, J. D., & Morris, H. C. (1982). *Solitons and nonlinear wave equations*. New York: Academic.
17. Drazin, P. G., & Johnson, R. S. (1989). *Solitons: An introduction*. Cambridge: Cambridge University Press.
18. Drazin, P. G., & Reid, W. H. (1981). *Hydrodynamic stability*. Cambridge: Cambridge University Press.
19. Edwards, C. M., Howison, S. D., Ockendon, H., & Ockendon, J. R. (2008). Non-classical shallow water flows. *IMA Journal of Applied Mathematics, 73*, 137–157.
20. Garabedian, P. R. (1964). *Partial differential equations*. New York: John Wiley and Sons.
21. Germain, P., & Lee, E. H. (1973). On shockwaves in elasto-plastic solids. *Journal of the Mechanics and Physics of Solids, 21*, 359–382.
22. Glass, I. I., & Sislan, J. P. (1994). *Nonstationary flows and shock waves*. Oxford: Oxford University Press.

© Springer Science+Business Media New York 2015
H. Ockendon, J.R. Ockendon, *Waves and Compressible Flow*,
Texts in Applied Mathematics 47, DOI 10.1007/978-1-4939-3381-5

23. Greenspan, H. P. (1968). *The theory of rotating fluids*. Cambridge: Cambridge University Press.
24. Guderley, K. G. (1962). *The theory of transonic flow*. Elmsford, NY: Pergamon. Translated by J. R. Moszynski.
25. Hayes, W., & Probstein, R. (1959). *Hypersonic flow*. New York: Academic Press.
26. Hinch, E. J. (1991). *Perturbation methods*. Cambridge: Cambridge University Press.
27. Hodges, C. H., & Woodhouse, J. (1986). Theories of noise and vibration transmission in complex structures. *Reports on Progress in Physics, 49*, 107–170.
28. Howell, P., Kozyreff, G., & Ockendon, J. (2009). *Applied solid mechanics*. Cambridge: Cambridge University Press.
29. Howell, P. D., Ockendon, H., Ockendon, J. R. (2012). Mathematical modelling of elastoplasticity at high stress. *Proceedings of the Royal Society A, 468*, 3842–3863.
30. Ilyushin, A. A. (1963). *Plasticity, foundations of the general mathematical theory*. Moscow: Moscow Academy of Sciences.
31. Kaouri, K., Allwright, D. J., Chapman, C. J. & Ockendon, J. R. (2008). Singularities of wavefields and sonic boom. *Wave Motion, 45*, 217–237.
32. Kevorkian, J., & Cole, J. D. (1981). *Perturbation methods in applied mathematics*. New York: Springer-Verlag.
33. Keyfitz, B.-L. (1999). Conservation laws, delta shocks and singular shocks. In *Nonlinear theory of generalised functions*. Research notes in mathematics (pp. 99–111). Boca Raton: Chapman and Hall.
34. Lax, P. D. (1953). Nonlinear hyperbolic equations. *Communications on Pure and Applied Mathematics, 6*, 231–258.
35. Leveque, R. J. (2004). The dynamics of pressureless dust clouds and delta waves. *Journal of Hyperbolic Differential Equations, 1*, 315–327.
36. Liepmann, H. W., & Roshko, A. (1957). *Elements of gas dynamics*. New York: John Wiley and Sons.
37. Lighthill, M. J. (1978). *Waves in fluids*. Cambridge: Cambridge University Press.
38. Mora, P. (2003). Plasma expansion into a vacuum. *Physical Review Letters, 90*, 185002.
39. Mouhot, C., & Villani C. (2010). Landau damping. *Journal of Mathematical Physics, 51*, 015204.
40. Newton, I. (1871). *Principia*. New York: Daniel Adee. Translated by A. Motte (1846).
41. Ockendon, H., & Ockendon, J. R. (1995). *Viscous flow*. Cambridge: Cambridge University Press.
42. Ockendon, H., & Tayler, A. B. (1983). *Inviscid fluid flows*. New York: Springer-Verlag.
43. Ockendon, J., Howison, S., Lacey, A., & Movchan, A. (1999). *Applied partial differential equations*. Oxford: Oxford University Press.
44. Ockendon, J. R., & Ockendon, H. (1973). Resonant surface waves. *Journal of Fluid Mechanics, 59*, 397–413.
45. Ockendon, J. R., & Ockendon, H. (2001). Nonlinearity in fluid resonances. *Meccanica, 36*, 297–321.
46. Ockendon, J. R., & Tew, R. H. (2012). Thin layer solutions of the Helmholtz and related equations. *SIAM Review, 54*, 3–51.
47. Perego, M., Howell, P. D., Gunzburger, M. D., Ockendon, J. R., & Allen, J. E. (2013). The expansion of a collisionless plasma into a plasma of lower density. *Physics of Plasmas, 20*, 052101.
48. Riley, N. (2001). Steady streaming. *The Annual Review of Fluid Mechanics, 33*, 43–65.
49. Russell, J. S. (1845). Report on waves, Rep. 143th Meet. British Association for the Advancement of Science, New York.
50. Santosa, F., & Symes, W. W. (1991). A dispersive effective medium for wave propagation in periodic composites. *SIAM Journal on Applied Mathematics, 51*, 984–1005.
51. Stewartson K. (1964). *The theory of laminar boundary layers in compressible fluids*. Oxford: Oxford University Press.
52. Van Dyke, M. (1975). *Perturbation methods in fluid dynamics*. Stanford, CA: Parabolic.
53. Van Dyke, M. (1982). *An album of fluid motion*. Stanford, CA: Parabolic.
54. Whitham, G. B. (1974). *Linear and nonlinear waves*. New York: Wiley.
55. Zemansky, M. W. (1968). *Heat and thermodynamics*. New York: McGraw Hill.

Index

© Springer Science+Business Media New York 2015
H. Ockendon, J.R. Ockendon, *Waves and Compressible Flow*,
Texts in Applied Mathematics 47, DOI 10.1007/978-1-4939-3381-5

Printed in the United States
By Bookmasters